Errata

Please note:
On the map, on pages 12–13, two red dots are missing, and the places and names for numbers 8 and 49–56 are not correctly shown on the map. On this corrected reproduction of the map, the dots are correctly located, and the numbers match the place names in the legend on page 12.

NATURAL
WONDERS

OF THE WORLD

ABBEVILLE PRESS PUBLISHERS

NEW YORK   LONDON   PARIS

9
Hawaii's Kilauea volcano erupts, an event integral to the formation of planet earth. Over the course of millions of years, volcanism has shaped the world we live in.

10–11
The north face of Mount Everest, the tallest mountain on the planet at 29,028 feet above sea level.

C O N T E N T S

1
The peaks of Kata Tjuta seem to glow a fiery red at sunrise and sunset, the reddish light of the sun accentuating the natural tint of the rock.

2–3
Antarctica supports such animal species as the chinstrap penguin, seen here on a gorgeous blue iceberg in the Weddell Sea.

4–5
Lying east of Serengeti and Ngorongoro Crater and west of Mount Kilimanjaro in Tanzania's Great Rift Valley, Lake Natron lies amid a wild landscape of the cones of ancient volcanoes.

6–7
A few of the many coral islets called motus that ring Bora Bora in French Polynesia are seen in this view.

8
The Atlantic Ocean crashes onto the beaches of Portugal's Algarve Coast, highlighting the beauties of the sea.

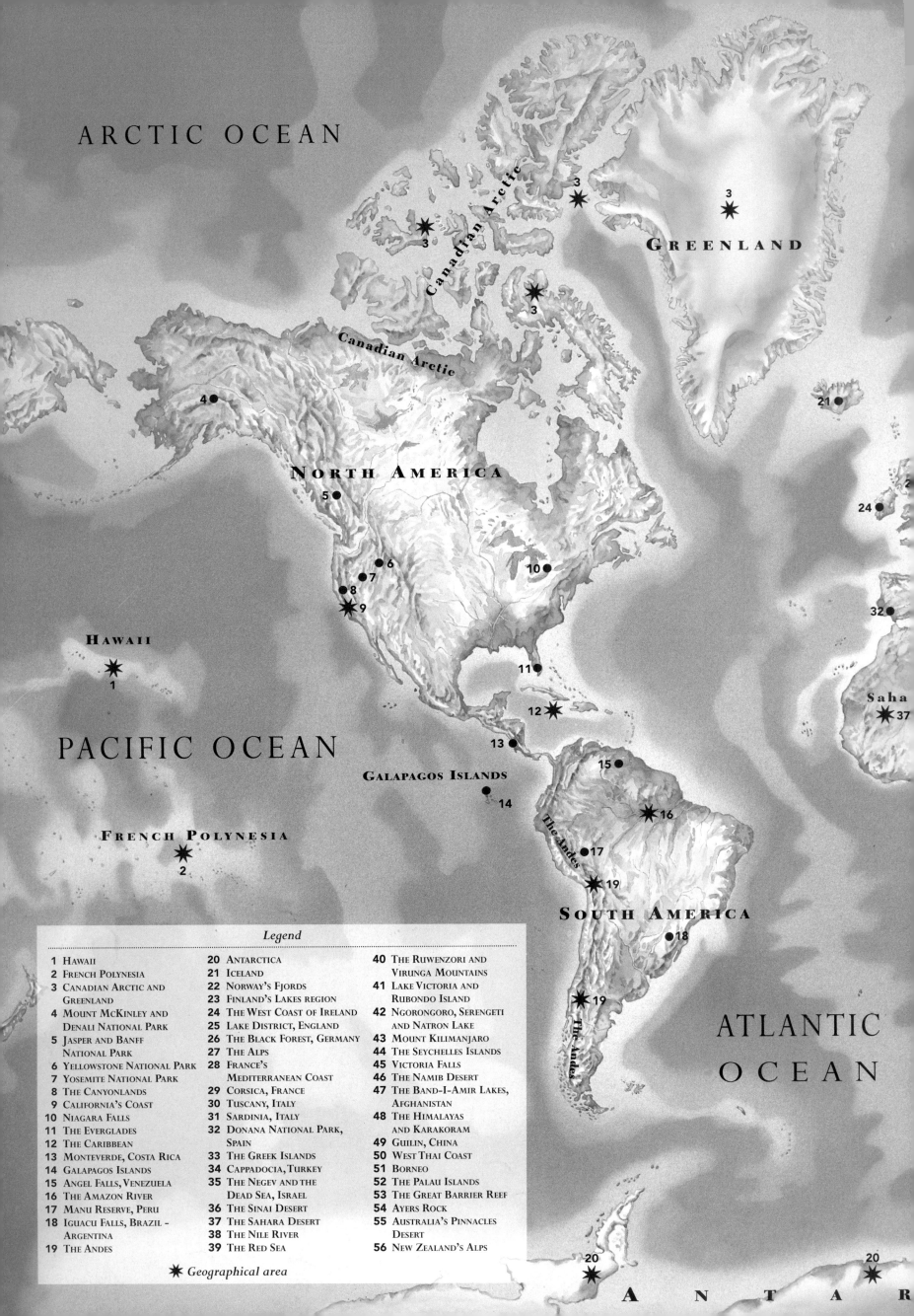

ARCTIC OCEAN

3

3

GREENLAND

3

3

3

Canadian Arctic

Canadian Arctic

21

4

24

NORTH AMERICA

5

6

7

8

9

10

32

HAWAII

1

11

Saha

12

37

PACIFIC OCEAN

13

GALAPAGOS ISLANDS

14

15

16

FRENCH POLYNESIA

The Andes

17

2

19

SOUTH AMERICA

18

ATLANTIC

OCEAN

19

The Andes

20

20

A N T A R

ARCTIC OCEAN

N

22 ✴
22 ✴
23 ●

26 ●
27 ●
30 ●
31 ●

EUROPE

ASIA

33 ●
34 ●

35 ●
36 ●

37 ✴
esert

38 ✴
39 ●

38 ✴

Africa

47 ●
48 ✴

48 ✴

PACIFIC OCEAN

40 ●
41 ●
42 ●
43 ●

45 ●

46 ●

SEYCHELLES
44 ●

50 ● THE PALAU ISLANDS

49 ●

Great Barrier Reef
51 ✴

53 ●
52 ●

AUSTRALIA

NEW
ZEALAND
54 ●

INDIAN OCEAN

20 ✴

C T I C A

14
California's Santa Cruz Island is about twenty-four miles long and comprises ninety-six square miles. The highest of the Channel Islands' mountains is found on Santa Cruz, standing 2,470 feet. Over six hundred plant species and 140 bird species live here. The cliffs along the shoreline, offshore rocks, and tidepools all provide breeding habitat.

15 Center
Set off against brilliant blue water, the legendary pink beach at Budelli, off the coast of Sardinia, is extraordinarily beautiful.

15 Bottom
The lovely flat seashore and beach near the Niger River delta fronts an area of thick tropical bush. This coastal belt eventually gives way to the broad, grassy savannas of the interior, where most of the nation's farming is done.

15 Top left
A storm pounds the shoreline of Camogli along the Ligurian coast of Italy. Water, necessary for all life on earth, is one of the forces that shapes and defines the romantic coastlines of continents and islands.

15 Top right
The beautiful Cayman Islands of the Caribbean market their lush tropical plant life, white sand beaches, and clear warm waters to lure visitors, especially from northern climates, during the winter months.

PREFACE

Our world is composed of many kinds of natural wonders. Some of these treasures are seen every day of our lives. We come to accept them and then begin to ignore them. Yet there are places on this earth that stretch the boundaries of human imagination in their immensity and visual splendor. These are the places that cannot be taken for granted, the true wonders of our world.

Some of these places are astounding because of their sheer physical beauty, like the green coast of Ireland, the high Rocky Mountains of Canada, or the green fields of Tuscany. Who can believe their eyes upon seeing the island paradise of the Seychelles, the coral wonders of the Great Barrier Reef, the perfect symmetry of Mount Fuji, or the plunging coastline of Norway's fjords?

Other places can be classified as natural wonders because they are incomprehensibly vast. Who can understand the majesty of Victoria Falls, the endless stretches of the Sahara Desert, the infinite twists and turns of the Nile River, or the icy altitudes of the Himalayas?

Still other places are natural wonders because they harbor ecosystems that are vanishing in our modern world. The national parks of Kenya, Tanzania, and Zaire in Africa; Yellowstone and Yosemite in North America; Monteverde in Central America; Manu in South America; the Galapagos Islands off South America's coast; and countless others all contribute to the preservation of little worlds, fragile and unique. Who can understand the multiplicity of the plant and animal life of the Amazon Rainforest; the incredible life forms that live in the world's driest and oldest desert, the Namib; or the importance of the preservation of the vast, thick ice caps of Antarctica and Greenland?

These natural wonders are defined by many criteria, but each is beautiful in its own way. All of them are difficult to

16–17
Glacial runoff from Mount Alpamayo in the South American Andes contributes to rivers that join to form the mighty Amazon. In turn, the Amazon River supports a rainforest whose plants and their expulsion of oxygen make it possible for the entire planet and all its creatures to live in a moderate temperature range and to breathe its atmosphere. Each of the earth's natural wonders in some way benefits the entire planet's ecosystem.

16 Bottom right
A wondrous landscape often includes features that remind human beings of more mundane things. Here is the rocky Alpine spire known as the Dente del Gigante, or the giant's tooth, which rises above the jagged ridgelines of Mont Blanc in Italy.

17 Top
Sometimes a geographical feature takes on a special significance, one beyond its physical form or beauty. Such is the case with Mount Fuji, which has become an important symbol for the Japanese nation and for its people.

16 Bottom left
Water is one of nature's primal forces, particularly in the transition from its frozen state to liquid. Without water like that shown in this view of Lake Moraine in Canada's Banff National Park our planet would be lifeless.

comprehend. Most are difficult to reach, and the majority of the world's people will be lucky to see even a fraction of them in a lifetime.

The natural wonders presented here are showcased in such a way that the reader can take a special journey around the globe. Each of the earth's most spectacular places may be enjoyed for its physical beauty or unique terrain, or for its plant and animal life, or its remoteness. Photographs and text reveal the marvels of each of the world's continents, the islands of the Pacific, the Arctic and the Antarctic, and they help to explain how such marvelous places were created and how the endure. And they can be enjoyed here without many of the worries that plague explorers—insect bites, snakes, avalanches, and even jetlag!

The sites presented are a curious combination of the rugged and the fragile. Despite the vastness of the Amazon, the Sahara Desert, the Himalayas, the North American canyon country, and the world's oceans and permanent ice fields, the presence of man can easily affect these areas. Machines and pollution have vast potential for damaging fragile ecosystems. The world's resources are undergoing great stresses and strains as we enter the twenty-first century. In a world with rapidly diminishing natural areas and growing populations, these incredible deserts, jungles, glaciers, beaches, volcanoes, waterfalls, and other natural features provide a special insight into regions still untouched by the hand of man. Today, as will be seen, national parks, conservation areas, and preserves created by people and nations of foresight throughout the world are struggling to protect and preserve our natural areas. In many cases only the enforced protection of natural wonders will ensure their preservation for generations to come.

This book is a portrait of our earth's most precious wonders. It celebrates the beauty of the world's elemental forces, the ancient primal materials of earth, wind, fire, water, and gravity. The earth has made itself over time with such processes as tectonic movement, volcanism, uplift, and erosion, and it continues to remake itself. Wind sculpts the exposed areas of the earth into fantastic shapes. Everything from rock to sand to waves to snow is arranged and rearranged by the earth's winds. The winds carry seeds and spores, some aided by birds and insects, to new lands and new continents. Fire lights the night skies of tropical islands, Andean peaks, and African rainforests as volcanoes spew forth red lava that takes away the old life it touches while creating new, fertile ground on which new life may germinate. And water, the essence of all life we know on earth, enables

plants, animals, and microscopic organisms to live. Water also sculpts the earth, with the moving ice of glaciers, the roaring torrents of rivers, the expanding power of ice in rock crevasses, or the pounding waves of the mighty oceans.

Without these elemental forces, our world would not be the same and these natural wonders would not exist. Add to them the natural rhythms of the earth's rotation, the pull of the moon, and the incalculable importance of the proximity of the sun, warming the earth to the proper temperature to sustain biological organisms, and you have the setting for their creation. Together, these forces constitute a humbling experience. Few can deny that a larger force is at work here,

a force discerned in the beauty of a snowflake or a mountain and its glacier, in a drop of water or the expanse of the entire Pacific Ocean, in a grain of sand or in the Sahara's massive dunes. It is a force as incomprehensible as it is beautiful. And in the end, these wonders of nature are a result of God's handiwork—the divine spark that set it all in motion, including our own human ability to perceive and appreciate them.

The dazzling photographs that follow reveal many of the world's most spectacular wonders. Cumulatively, these profiles provide an incredible overview of the world we live in and the power and majesty of the natural forces that created and continue to shape it.

18 Top left
Badlands National Park in South Dakota includes wild and fascinating rock formations such as these, rising from the flat prairie and stretching for miles in many directions.

18 Top right and bottom right
The Rufiji River and its delta are seen in these views of western Africa. The Rufiji is located in Tanzania, reaching the Indian Ocean south of Dar es Salaam.

18 Bottom left
Another beautiful case of sculptural rock, this time volcanic tufa located in Uchisar in the Cappadocia region of central Turkey. The huge formations were crafted by the forces of nature, then crafted further by the hand of man.

19
The sinuous forms of wind-formed sand dunes in the Sahara Desert can look more like a work of art than a real scene, even standing before them. Perhaps that is a true test of whether or not a feature qualifies as a wonder of nature: Can viewers be certain that what they are seeing is real?

20–21
Even in a place of seeming desolation, life somehow manages to gain a foothold and maintain it. Here, in the dry heat of the Namib Desert amid nothing but tall dunes of sand, plants and animals live on the cold mists blown in each morning off the Atlantic Ocean and the brief seasonal flooding of dry riverbeds inland.

22–23
The majestic expanse of Bryce Canyon's red hoodoos covered with snow reveals the contrasts found within the beauty of the desert regions of the United States. Wind, water, and ancient oceans sculpted this incredible region.

EUROPE

EUROPE

INTRODUCTION

Great peninsulas jut from her storm-pounded coasts in the north, while other peninsulas in the south huddle about a warm, sun-soaked sea. Her coastlines are long and graceful, while in the center she rises with gorgeous snow-capped mountains. Island gems surrounded by azure waters lie near her shores. Her lakes gleam like glass, and trees line the banks of her rivers, which coarse to the sea. Her appealing climate creates lush forests and productive farms. She is Europe, temperate cradle of Western civilization.

The continent was named after Europa, who in Greek mythology was the daughter of Phoenix. In Europe politics, religion, philosophy, science, art, drama, and music reached great heights. Due to the expansive nature of European society and the penchant of her people for exploration and trade, the civilization they pioneered was exported to the rest of the world. Colonization accelerated the pace of this dissemination of knowledge, until by the early twentieth century most of the globe was under the military or cultural sway of at least one European country.

Some of the most beautiful and memorable places on Earth are in Europe. The Alps, the Riviera, the Norwegian fjords, the Greek islands, and the Black Forest spring immediately to mind, while many other natural wonders, perhaps not as well known, can be found, like those in Iceland, Finland, Spain, Italy, England, and Ireland. Europe's geographical features are exceptionally beautiful. Although many other places in the world are called alps or a Riviera, many find them just pale imitations of the originals that are located in Europe.]

Europe is one of the smallest continents, covering just 4,063,000 square miles, with a coastline of 37,887 miles. It is four thousand miles long and three thousand miles wide, and with 728 million people has the second largest population of any continent. Europe is actually a large peninsula that joins Asia at the Ural Mountains in Russia. Some geologists do not consider Europe to be a separate continent at all, and put Europe and Asia together as one large continent they call Eurasia.

Urban areas in Europe have their own special charms. Many cities are famous because of their magnificent art and architecture. Other are manufacturing centers of great importance. But Europe is made up of much more. Its incredible geographical features are beautiful in their own right, but also central in giving Europe its important place in world history. Its rich soil has made it possible to feed its people. Its forests, water power, and coal and iron deposits have created the power necessary for industry and technology. And its predominantly temperate climate has made it possible to produce food and the materials necessary for housing. Throughout history, these gifts have given Europe's men and women the leisure time necessary for intellectual and scientific advancement.

25 Bottom left
The rolling foothills of the Dolomite Mountains spread forth in a delightful green carpet of meadows in the Alto Adige area of Italy. The Val Gardena, seen here, is a major skiing, rock climbing and mountaineering center near Bressanone.

25 Bottom right
Looking east to Lake Lucerne from Mount Pilatus, one can see the visual glories of central Switzerland in the heart of the Alps. This storied and romantic area inspired Richard Wagner, Goethe, Mark Twain, and even Britain's Queen Victoria, who scrambled to this very spot in 1868.

24–25
The scenic Alps above the Val d'Aosta region of northern Italy exemplify the European highlands. From Mont Blanc to the Matterhorn and Monte Rosa, this chain of rugged, majestic mountains and glaciers forms what is known as The Roof of Europe.

26–27
The gray stone mass of Scotland's Eilean Donan Castle strategically sits at the confluence of several lochs in northwestern Scotland. Lochs—deep glacial lakes and inlets similar to fjords—punctuate the northern portion of Scotland. The lovely colors of the trees and grasses is set off by the cloudy skies, which are pierced intermittently by sunlight.

Europe's place on the Gulf Stream contributes to its temperate climate. The Stream sweeps warm ocean currents from the Caribbean Sea across the Atlantic to Europe's shores. The warm westerly winds that blow in off the Atlantic Ocean also help temper the climate. Arid regions exist only in the southeastern regions of the continent within Russia. Most of the continent is cool and moist, with a temperate to cold climate and plenty of rainfall, but on the south side of the Alps the Mediterranean region is normally hot, dry and sunny.

Europe has four major land regions. The first consists of the northwest highlands of Ireland, northern France, northern Great Britain, Norway, Sweden, and northern Finland. Here thin soil makes for poor farming. Lying near the surface of the earth is the rocky ancient bedrock known as the Fenno-Scandian Shield, which was formed in the Precambrian age. During several successive ice ages, glaciation carved deep valleys, lakes, and fjords in the region. This contributes to the gorgeous undulations of the predominantly green landscape and to the deep rocky inlets found along the coastlines. On the other extreme is Iceland. There the planet's inner forces break through the surface in the form of dramatic volcanoes and hot springs.

Other sites show the variety of riches the region encompasses. The ocean laps at the towering cliffs of Norway's fjords, whose inlets become placid rivers as they wind further inland. Finland has thousands of famed lakes that were scooped out by glaciation. Prized as recreational spots, they form patches of brilliant blue strung out like pearls on a green velvet cloth. On Ireland's west coast, storms kick up wind and waves against stubborn green cliffs, while England's beautiful Lake District, also carved by glaciers, is nestled snugly in the nation's northwest corner.

Europe's central plains make up its second land area. This is the region where most Europeans live. Extending across

26 Bottom left
Norway's Lofoten Islands Fjord near Lingvaer demonstrates the quiet beauty of the region in the winter. These deep inlets, sheltered from the ocean waters, punctuate the entire west coast of Norway.

27
The sheer cliffs of Kilt Rock on western Scotland's Isle of Skye overlook the Sea of Hebrides. Temperate plants, even magnolias, bloom in this Siberian latitude, thanks to the warm waters of the Gulf Stream.

26 Bottom right
The striking white limestone cliffs of France's Normandy coast near Etretat rise suddenly from the English Channel about one hundred miles northwest of Paris. Peaceful villages dot the fertile lands above the cliffs, protective bastions that guard the coastline.

INTRODUCTION

28 Top left
*Although its appearance may be
forbidding, the Italian island of
Ponza is a delightful spot. White
cliffs hide fragrant flowers,
vineyards, grottoes, and beaches.*

28 Top right
*A view of Kekova Beach, one of
thousands of similar spots along
Turkey's Mediterranean Coast.
This gorgeous region is set off by
the blue waters of the sea.*

INTRODUCTION

most of the continent from west to east, the plains are formed from sedimentary material and are Europe's most fertile area, with wind-blown topsoil called loess added to the already rich soil. The plains, which rarely rise 500 feet above sea level, begin in southeastern England and run across France, Belgium, the Netherlands, northern Germany, and Poland to the Ural Mountains in Russia. Their beauty ranges from green farm fields marked off by hedgerows in the west to flat plains of golden grain waving in the east.

Mountain system, which runs across southern Europe. In geological time Alps are relatively young mountains. Beginning in Spain, the range includes the Sierra Nevada, Pyrenees, Apennines, Carpathians, Balkans, and the Caucasus, as well as the Alps themselves. Surprisingly, the highest mountain in the system is not in Switzerland or France—it is the 18,481-foot Mount Elbrus in the Caucasus range of southern Russia. The Alps and their foothills cover the Mediterranean area of the continent as well. The

The central uplands form the third of Europe's land areas. They run from Portugal through central Spain, France, and southern Germany to the Czech Republic. Low-lying areas meld with separate plateaus, hills, and mountains, which range from five hundred to four thousand feet above sea level. Dense forests and open grasslands cover most of the region, which is composed of mixed geological components created by folding, faulting, volcanoes, and uplift. The beautiful windswept plateaus of Spain are also in this region, as are the green-clad Jura Mountains that run between France and Switzerland and the incredible Black Forest of Germany.

The last of the four European land regions is the Alpine

gorgeous, towering, glacier-clad mountains mix with sunny coastlines along the Côte d'Azur in France, Sardinia and the Greek Islands. Foothills were never more beautiful than those found in sun-drenched Tuscany, while Spain's Doñana National Park preserves an example of the coastal marshlands of the region.

The continent of Europe presents landscapes of great beauty. And they are useful as well, with a temperate climate and lengthy growing season. The hospitable region nurtured cultures that spread throughout the world, changing life on the planet forever and making it what it is today. But Europe's natural wonders alone make it one of the most beautiful places on earth.

28 Bottom
Located near the town of Vieste on the spur above the heel of Italy's boot, the Pizzomunno is a stunning landmark set in the midst of gorgeous scenery. The limestone pillar is on the east coast of Italy along the Adriatic Sea, and, according to legend, it represents a fisherman who turned to stone after his lover drowned.

28–29
The waters are calm and the sun is high in this scene of Rabbit Beach on the island of Lampedusa. Covering only eight square miles, Lampedusa is located south of Sicily and is actually closer to Africa than Italy. The island was not inhabited until 1843, when Ferdinand of Bourbon encouraged fishermen to immigrate.

29 Bottom left
Punta di Capo is the final promontory of the Portofino peninsula along the gorgeous Italian Riviera. Portofino is located twenty-five miles southeast of Genoa on the Ligurian Sea. It has become a desirable location for vacationing, and the old houses and villas are extremely expensive to rent or buy.

29 Bottom right
Porto Greco along the Sardinian coastline speaks to the romance of this special island. Clear blue waters, which often become a brilliant and startling shade of turquoise, invite swimming, snorkeling, and diving offshore. On land the rocky topography is tempered by Mediterranean vegetation and by brightly colored wildflowers.

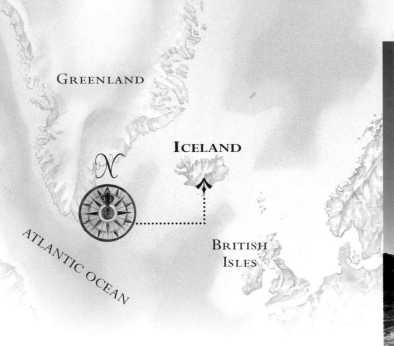

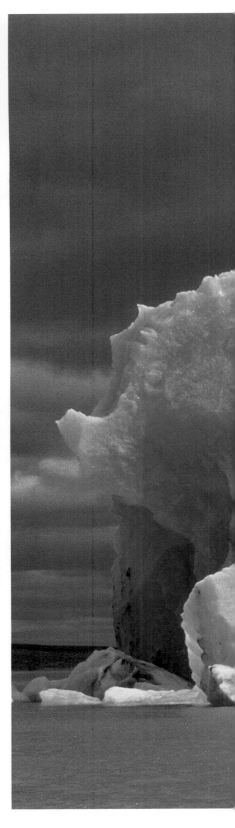

LAND OF FROST AND FIRE

ICELAND

Imagine an otherworldly place in the North Atlantic rent by faults and fissures, covered with glaciers and active volcanoes, spouting geysers and hot springs. It is the island nation of Iceland, whose black lava, grassy green valleys, yellow sulfur rocks, and hot geysers make it a colorful land of wonder. Streams flow from the interior, where waterfalls gush and a lunarlike landscape supports few trees.

Iceland has long seemed more like a land sprung from the world of science fiction than a real place. In Jules Verne's 1864 novel, *Journey to the Center of the Earth*, Iceland was the entrance to a system of underground caverns leading to the inner recesses of the planet. Although no easy access to the earth's interior actually exists, Iceland is a place where scientists can readily study the geological forces that make up our planet.

The wonders on exhibit in Iceland are the result of plate tectonics. The earth's seemingly solid surface is actually a restless jigsaw puzzle of slabs known as tectonic plates. These pieces of the crust and mantle ride on a dense layer of the earth's mantle called the asthenosphere, and they move about and make contact with one another. Where the plates meet, there are faults, and they are the sites of frequent earthquakes. Many fault lines also show the results of the collision of plates, resulting in volcanoes and geothermal activity like hot springs and geysers. Iceland is a special place because it exhibits the results of two ocean plates drifting apart.

30 Bottom
Snowmelt from the Gigvjokull, one of Iceland's smaller glaciers, is seen wending its way to the sea near Thorsmork. The incredible nature of glaciers and their ability to scour and sculpt the earth is studied by geologists in Iceland.

30–31
A towering iceberg calved from the Vatna Glacier floats in the cold waters surrounding Iceland. The serene, majestic beauty of icebergs belies the danger they present to ships.

31 Bottom
In addition to ice, snow, and glaciers, Iceland knows the heat of the earth as well. This fumarole is one of hundreds of such vents in Iceland, where the movement of tectonic plates allows the earth's heat to escape from the ground. Reykjavik, the capital of Iceland, along with several of the nation's other cities, is heated and powered by geothermal means alone.

Iceland is located in the North Atlantic about 185 miles east of Greenland and 620 miles west of Norway. It is the most sparsely populated country in Europe, with an average of just seven inhabitants per square mile. Nearly four-fifths of the country is uninhabited. The 3,700-mile-long coastal zone is where most of Iceland's inhabitants live. Although the island is close to the Arctic Circle, the coastal regions are warmed by the Gulf Stream, which keeps the temperature moderate and consistent year-round. Parts of the coastline are made up of brown cliffs. In other areas, villages and cities huddle near the water on coastal plains backed by blue mountains, partially covered with snow. The cloudy, often rainy weather and the cool temperatures create a somber mood, filled with quiet beauty.

Iceland is a land of water, with rivers flowing swiftly down from the glaciers on the central plateau, creating dramatic waterfalls like this one.

The path of seasonal glacial runoff at Torsmork forms a river of green. Iceland's unusual combination of features makes it a place unlike any other on earth.

32 Bottom
This waterfall, Selijaland Foss, shows how dramatically the plateau lands drop to the surrounding coastal plain in Iceland. The brilliant hues of the countryside combine with the powerful beauty of the waterfall, revealing yet another aspect of Iceland's marvelous landscape.

ICELAND

Iceland has more than its share of unusual and startling features, most of which are clustered in an uninhabitable central plateau that makes up the bulk of the 39,769-square-mile island. The solid lava tableland of this region, which is rugged and barren, lies about 2,500 feet above sea level. The 6,952-foot mountain called Hvannadalshnukur is the island's highest peak.

A line of faults runs across the island, making the plateau a land of violent natural wonders, where earthquakes are common. The island itself was formed by volcanic activity, which continues today: There are more than one hundred volcanoes, craters, thermal springs, and geysers covering the landscape. About twenty of Iceland's volcanoes have erupted in recorded history, the most recent eruptions occurring on Mount Hekla in 1947, 1980 and 2000.

In addition to volcanoes, Iceland has more hot springs and sulfur stream beds than any other country. The very word geyser comes from the name of Iceland's most famous hot spring, Geysir. Located in Haukadalur in southern Iceland, Geysir spouts a column of water 180 feet high every five to thirty-six hours. Strokkur, another geyser in the same field, spouts every few minutes. These geothermal fields are brown and gray, with steam vapors rising from fumaroles and from bubbling, sulfurous springs of mud that snap and pop. The colorless terrain and the ferocious nature of these geothermal areas combine to give the plateau the forbidding look that

32–33
Iceland is a land of constant surprises and beauty, epitomized by Gull Foss, the Silver Waterfall, in the southwestern part of the island. Gull Foss is on the Hyita River, which is formed from snowmelt from the Lang Glacier. This dramatic waterfall is less than fifty miles from Reykjavik.

34–35
The enormous Vatna Glacier of southeastern Iceland can be seen from the coast in this view. Glaciers cover nearly 12 percent of the island nation, with the Vatna Glacier alone encompassing 3,240 square miles. Offshore, the glacial melt turns to icebergs that glow with a blue sheen.

Verne conjured up in *Journey to the Center of the Earth*.

Iceland has at least 250 low-temperature geothermal areas with a total of about 800 hot springs. The average temperature of the hot springs is about 170 degrees Fahrenheit, and these hot springs are a potential source of natural energy and power. Deildartunguhver, the largest hot spring in Iceland, has a flow of forty gallons of boiling water per second, while the Torfa Hot Springs alone generate power equal to 1,500 megawatts.

Contrasting sharply with the heat of the island's vast geothermal activity, Iceland has more than 120 frigid glaciers, which cover more than 4,500 square miles, or nearly 12 percent of the country. The contrast of the hot springs and white, icy glaciers can be startling, but the island sports all sorts of surprising features. Avalanches are common in the north, northwest, and east. Like glaciers world wide, Iceland's have been shrinking and retreating in recent years. Its largest glacier, Vatna, in the southeastern part of the country, covers 3,240 square miles, is 3,000 feet thick, and is equal in size to all the remaining European glaciers put together. This huge region of slowly crawling ice often reflects the aurora borealis, a strange atmospheric effect that produces waving bands of color in the night sky.

The majestic glaciers, the fabulous hot springs, and the eery glow of the aurora on the wind-blown tableland combine in Iceland to create one of the world's most unusual landscapes. It is a visually dazzling country, one filled with fascinating geological phenomena, a place little known to most, a land of frost and fire filled with an incredible number of natural wonders.

TOWERING MAJESTY

NORWAY'S FJORDS

Long, narrow fingers of ocean penetrate the coastline of Norway. Called fjords, these inlets are renowned for their physical beauty and because they are fine harbors. Snow-clad mountains are reflected in the pale green waters, while bright green fields contrast with the rugged grays of the granite walls, where waterfalls tumble down to the ocean. Fjords are beautiful, lively, and awe inspiring, and there are plenty of them to marvel at and explore. The Norwegian coast runs for about 1,650 miles, but taking into account all of the fjords and peninsulas, the full length of the coast is 12,500 miles, or about half the circumference of the earth.

Fjords are glacially deepened inlets with sheer, high sides. They have a U-shaped cross section and a submerged shelf on the seaward side composed of moraine detritus. Fjords occur primarily in southern Alaska, British Columbia, southern Chile, Greenland, the south island of New Zealand, and Norway. The moving ice of glaciers created the fjords by entering river valleys millions of years ago. A glacier is created when annual snowfall accumulates for a series of years without melting during the summer months. As a glacier enters a mountain valley it acquires loose material from stones, rocks, and even boulders. As the glacier continues to move, this loose material acts like a carving tool, gradually whittling and eroding landforms into new shapes. A glacier widens, deepens, and straightens a mountain valley, changing a V shaped cross section into the U shape of a glacial trough. In the case of the fjords, the ocean spilled into the glacial valley after the ice retreated, filling it and creating a natural harbor.

Most of Norway's fjords have steep, rocky, forested walls. The warm waters of the Gulf Stream help to keep coastal temperatures moderate keeping most fjords ice free year round. The longest fjord, Sognefjorden, winds inland for 120 miles, with many branches that jut from the main channel. Fjords are not only bordered by steep walls. They are deep channels, which hold a great deal of water, making it possible for large ocean-going ships to penetrate far into the interior. Remnants of debris left by the glaciers that formed them make fjords shallower at their mouth than they are inland, where the water in fjords can be very deep. For example, parts of Sognefjorden are 4,291 feet deep. Mountains tower above it, more than 5,000 feet high on the average.

Today, fishing vessels and tour boats come and go from Norway's fjords, recalling the history of these famous waterways. For a three-hundred-year period, from 800 to about 1100 A.D., fierce sea raiders called Vikings sailed from the Norwegian coast attacking and pillaging communities in England and other parts of Europe. The Vikings also established colonies in Iceland, Greenland, and North America. Today the legacy of these master seamen and feared warriors is respected by the people of Norway.

The fjords along the Norwegian coast are incredibly varied. Some of the most breathtaking views may be found just north of Bergen, which is known as The Gateway to the Fjords. Such fjords as the Isefjord in north Siaelland are wide and dotted with islands, while the narrow Lysefjord, near Stavanger, is lined on both sides with walls of sheer rock up to 1,968 feet high. Hardangerfjord has gentle slopes covered with fields and orchards that allow vast, open vistas that sweep toward the incredible snow-capped peaks in the distance. White sheep graze in the fields bordering the placid blue-green water. A branch of Storfjord, Geirangerfjord, is particularly spectacular, with water curving through the interior of the country between almost perpendicular walls of dark gray rock, which are partly covered with a coat of thick, wild vegetation. Glittering ribbons of water stream from down the sides of the rock face, creating rainbows.

Today countless tourists travel to Norway to explore the beauty and the majesty of the country's famous fjords, entranced by the dramatic contrast between the majestic mountains and the clear blue water.

36 Top
Placid farms and quiet villages are often nestled throughout the hills surrounding fjords. Glaciers on the distant mountains seen here spill their meltwater downstream, creating spectacular waterfalls that plunge into the long, twisting fjord.

36 Bottom
Gorgeous, towering peaks rise above the green fields and blue waters of a Norwegian fjord. These long fingers of deep water lie just off the rugged Norwegian coast. In addition to the scenery, fishing is an attraction, with salmon available in the rivers, trout in the glacial lakes, and saltwater fish in the fjords.

36–37
A cruise ship winds its way out of Geiranger Fjord in south central Norway, a popular tourist destination. Fjords are perfect for large cruise ships, as their deep water allows them to steam far into the interior, with enough space to turn around.

37 Bottom left
The towering heights of Prekelstolen Rock are a perfect vantage point for viewing the beautiful fjord below.

37 Bottom right
Kvaenangen Fjord in northern Norway is located two hundred miles above the Arctic Circle. The fjord is a haven for Arctic wildlife.

38-39
The frozen expanses of Finland's lakes stretch across a countryside dotted with small hamlets. Scattered throughout are houses *made of wood and of logs. Winter sports on the frozen lakes are particularly popular and include skating, hockey, sledding, and ice fishing.*

38 Bottom left
Howling winds blow across the thousands of lakes in Finland's southern region. The lakes are more than a scenic wonderland. They serve as a transportation corridor and are a sporting paradise.

38 Bottom right
Rivers and canals course through the district, connecting many of the lakes. The calm, magical sound of water is everywhere, as it ripples over rocks and tumbles toward the sea.

LAND OF THOUSANDS OF LAKES

FINLAND'S LAKES REGION

Finns often refer to their country as The Land of Thousands of Lakes. In fact, an innumerable number of lakes dot the countryside, covering about 11 percent of Finland's territory. Estimates of the exact number range from 60,000 to nearly 200,000. During the last Ice Age, in the Pleistocene Epoch that ended about ten thousand years ago, glaciers scooped out lake basins in what is now Finland and Russia. The largest lake is the 680-square-mile Saimaa. Most of the lakes are clustered in the southeastern part of Finland in a lake district like few in the world.

Finland's lacelike pattern of lakes constitute an important resource. Connected by some 3,000 miles of streams and canals, they form an important transportation route. The Saimaa Lake system alone runs for 186 miles, eventually connecting to the Gulf of Finland. The lakes form waterways over which lumbermen float pine, spruce, and fir trees to market. With forested areas covering two-thirds of Finland, the lumber industry is vitally important to the nation's economy. In addition, the many streams leading from the lakes contribute to hydroelectric power.

People enjoy the peaceful majesty of the lake district on a regular basis, cooling off in its waters, fishing, hiking, and skiing amid the scenic grandeur. The lakes provide an escape not only for natives of the country, but for its many visitors. The entire area is covered with tall, stately pines. Lakes like Päijänne, Suvasvesi and Puulavesi have highly indented shorelines and sheltered bays and inlets that are backed by rocky cliffs or gently rolling farmland. Their natural beauty aside, the lakes are especially impressive because of their huge number. When seen from the air they seem to stretch into infinity.

Large gray boulders and felled timber are found in the wild and rugged land bordering the lakes. A labyrinth of dirt and gravel logging roads provide transportation throughout the region. And because the roads are rough, there is very little traffic and what there is moves slowly.

The lakes change their mood as the weather and the seasons change. In the summer the sweet smell of pine fills the woods, while autumn brings gray skies that contrast with the brilliant fall colors of birch, oak, maple and ash leaves. In

39 Top
Two female reindeer forage for grasses, leaves, moss, and lichens beneath the snowcover of Finland's lakes district. Reindeer are the only deer species whose females grow antlers. Reindeer in harness have been trained to pull sleds.

39 Bottom
A brilliant sky accents the dark green conifer trees bordering the lakes and streams of Finland's lake district. Quiet solitude rewards the hiker, snowshoer, or skier who ventures out into this lovely region.

40–41
On the southern coast of Finland, just west of Helsinki, the beautiful Ekenas Archipelago National Park leads from the inland lakes to the coastal islands. The park is gorgeous in all seasons, combining the rustic beauty of the interior with ocean views of the rocky coast.

40 Bottom
Tidal estuaries in Ekenas Archipelago National Park flow beside peaceful farms and cottages. The nation's small population and its rustic roads make it possible for everyone who visits the country to savor it without crowds and traffic.

41 Top left
Lake Päijänne is one of Finland's longest lakes, running one hundred miles through the south-central heartland, from Lahti on the south to Jyvaskyla on the north. The beautiful, calm waters of these glacial lakes are home to many popular sport fish.

41 Top right
Log cabins dot the landscape of the Finnish lakes, rising above the placid waters. Blending with the rustic scenery of golden fields, evergreen trees, and blue waters, the cabins house a proud people who appreciate the beauty and bounty nature has provided.

41 Bottom
Oulanka National Park lies in central Finland on the eastern border with Russia and includes the winding Oulanka Joki River. The river flows from the Finnish lakes eastward into Russia.

FINLAND'S LAKES REGION

winter a layer of brittle white snow covers everything, silhouetting the dark outlines of the trees against the lakes and crunching and crackling as people and animals move about. As trees begin to bud in the spring, a wealth of wildflowers grow in the nearby meadows. All these moods are reflected in the smooth, mirrorlike surface of the waters. The lakes themselves are deep and crystal blue, often frozen over with ice in the winter months. They are cold and clear and they team with such fish as perch, salmon, trout, and pike.

Many of the lakes are lined with homes and lodges built of wood, some of them made with logs. A favorite recreational pleasure of Finland is the sauna, and the ritual becomes something special when there is a cold, pristine lake nearby. A sauna is a room with a *kiuas* (stove) and *lauteet* (wooden benches or platforms). In the sauna, a steambath is created by pouring water over hot stones. Sauna lovers often massage themselves with whisklike tree branches, staying in the sauna until the heat is almost unbearable, when they dash from the sauna to jump into one of the clear, frigid lakes. The contrast between the heat and the cold is invigorating and a delightful (and extraordinarily relaxing) way to end a strenuous day of hiking or skiing.

Finland's sparse population of just five million is small for a European country and as a result the lake district is still wild and unpolluted. Evergreen forests look down upon crystal blue waters, while the only sounds are the birds singing, the wind in the trees, and the waves lapping at the shore. The lake district is a perfect place for being alone with nature, for fishing, canoeing, or just sitting in silence on a quiet beach savoring the area's serene beauty.

SCOTLAND

NORTH
SEA

ATLANTIC OCEAN

N

IRELAND

ENGLAND

42 Bottom
One of the smallest of the sixteen major lakes in the park, Crummock Water Lake is on the west side of the reserved district, barely eight miles from the Irish Sea. Its lush, fairly treeless scenery is typical of the region.

"A BLENDED HOLINESS OF EARTH AND SKY"

ENGLAND'S LAKE DISTRICT

It was the English poet William Wordsworth who once called the Lake District "a blended holiness of earth and sky." A resident of the area, Wordsworth championed the link between man and nature, using the Lake District as the epitome of that connection. His poetry publicized the area in the early nineteenth century and made it attractive to writers, poets, and artists. In 1802 Wordsworth wrote: "I wandered lonely as a cloud / That floats on high o'er vales and hills, / When all at once I saw a crowd, / A host, of golden daffodils." Today one would be hard-pressed to find such solitude, since the Lake District, which is a British National Park, attracts more than twelve million tourists each year.

The lakes and the rolling hills of the region are the legacy of the last Ice Age, when glaciers sculpted them from rock. The long, narrow lakes of the region are characteristic of glaciated valleys with end moraines at their mouths. Today the entire region is covered with the green of farm fields and forests, and the blue or gray of the water. Although the

mountains are not high nor the lakes large by world standards, the land and water sparkle with fragile beauty. And despite its generally low elevation, the Lake District is the site of England's highest mountain, 3,210 foot Scafell Pike.

Stone Age farmers once lived here, succeeded by Celtic peoples, then the invading Norsemen. All gradually cleared the Lake District's ancient forests, creating lush, parklike pastures. Today, the Norse words of the Vikings linger on in local terms. Dale means valley, a fell is a mountain or a hill, becks are brooks and a tarn is a high mountain lake. Tradition is strong in the region, but is being overwhelmed by the large numbers of tourists and recreationists who visit each year, all hoping to get closer to nature as Wordsworth once urged.

To protect the area, the Lake District National Park was created in 1951. Unlike park areas in some nations, much of the land within the park boundaries is privately owned. The park includes sixteen major lakes, including Ullswater, Bassenthwaite, Wast Water, Coniston Water, and the largest, Windermere, ten miles long and in most places no more than half a mile wide. The park's total area is 885 square miles and is the largest of the ten national parks in England and Wales.

Autos on the roads and hikers in the hills come to enjoy the Lake District's peaks, moors, hills, lakes, and soft green valleys lined with rock walls. For England this is a dramatic landscape, rare and vulnerable. The park is composed of private lands, mostly sheep farms, which are accessed through 1,800 miles of public trails and bridle paths. The local population of 40,000 people lives on the farms and in the small towns along the lakes.

The rugged hills are furrowed with rolling waves of earth, all covered with green and brown grasses. The placid blue lakes lie among these hills, which seem to rush precipitously into the rippling water. Water birds, particularly ducks, make this world their home. The green of the hills surrounding the waters seems almost iridescent, and the dramatic scenery is at once dramatic yet peaceful and tranquil.

People love hiking in this area of spectacular scenery. The sport of fell-running—that is, running up and down the region's mountains and hills— is popular. During such a run, or a nice, leisurely walk in the country, the visitor can enjoy the dappled sunlight and the shadows of clouds as they crawl over the face of the rolling green hills. As the sun glints off a

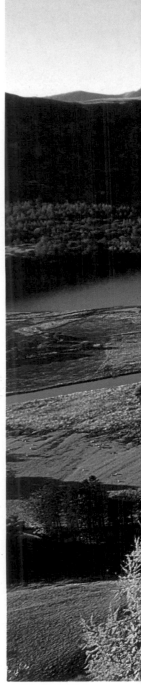

42–43
The gorgeous, calm reaches of Derwent Water, one of the major lakes in Cumbria's Lake District National Park, is seen *here from a point known as Surprise View. Motorized watercraft are banned on this lake, helping to preserve its serenity.*

lake or a waterfall, it is easy to understand how a spectacular landscape can inspire artists and poets to glorious heights.

It is fitting that one writer began the public interest in the Lake District and another worked to preserve it. Beatrix Potter, author of the Peter Rabbit stories and a resident of the area, worked to save the its fragile resources. When she died in 1943, she left fifteen farms and other properties to the National Trust, an act that inspired the establishment of the National Park in 1951. The National Trust today owns 140,000 acres or one quarter of the Lake District, including eighty-five farms, and fells, lakes, and tarns.

43 Top left
Probably the most famous single lake in the district, and certainly the longest, Lake Windermere is located in the southeastern corner of the 885-square-mile National Park. Towns along its eastern shore serve as a center for regional tourism.

43 Top right
One of the long, narrow lakes of the National Park, Ullswater is bordered by Glencoyne Wood and the Cumbrian Mountains, a combination that provides spectacular scenery. Most of the park consists of pastureland rather than forest. The trees were cleared from the fields long ago by the early Celtic peoples.

EMERALD
PENINSULAS

THE WEST COAST OF IRELAND

44 Top right
The green-mantled cliffs of a hidden bay rise above the Atlantic, sheltering a fine, sandy beach. This dramatic landscape on the northwestern coast of Ireland in Donegal County invites one to walk through and listen to its many wonders. Seabirds

like storm petrels, guillemots, puffins, razorbills, and Manx shearwaters glide above the turbulent coastline, searching for food. At low tide, the tidepools offer the seabirds a succulent banquet of creatures left behind by the retreating waters.

From a boat bobbing on the sea the Irish coast looks like a massive wall of stratified rock. From the air, it looks like a series of bony green fingers projecting into the ocean. From the ground, the green fields are bordered by rock walls that divide the land into irregular sections, rising and falling over hill and dale. Earth, sky, and ocean meet in incredible harmony on the gorgeous west coast of Ireland, the land of emerald peninsulas.

Because the coast is punctuated by many inlets and bays, no part of the country is more than seventy miles from the ocean. Two large bays on the west coast, Galway Bay and the mouth of the River Shannon, provide harbors for the ports of Galway and Limerick. Inland, the Kerry Mountains rise to dominate the southern peninsulas. Offshore, hundreds of small islands have been torn from the mainland over a succession of years by the powerful Atlantic Ocean. Beneath the water, there are deep valleys that testify to the power of nature.

The west coast of Ireland is composed of peninsulas and bays called rias, drowned river valleys that form long, funnel-shaped and branching inlets that meet the sea at right angles. Rias are found not only in Ireland, but also in southwestern England, northwestern France, and northwestern Spain. Unlike fjords, which are U-shaped, rias have a V-shaped profile. From above, Ireland's spiked coastline looks entirely different than a coastline composed of fjords.

Echoes of Ireland's past are found along the west coast. Humans settled the region by at least 3700 B.C. and built monuments and fortifications that stand to this day. Ancient forts and castles, fifth-century Ogham stones incised with the earliest form of Irish writing, and ancient monastic sites are not only picturesque. They also provide fascinating glimpses into the area's history. At one time the west coast of Ireland was literally thought to be the end of the earth, although local lore says that St. Brendan sallied forth from this coast in a coracle and made landfall in North America centuries before Columbus did in 1492.

Sandy beaches, green fields leading up to ocean cliffs, spectacular waterfalls, rock formations (more than four hundred million years old), caves, and arches line the coast. The ground is coated with a covering of furze, whins, and

44 Bottom and 44–45
A gentle ocean washes up on the Irish coast at Sligo Bay. Located just south of Donegal County, Sligo's dramatic peninsulas and bays are mantled with green, revealing the coastline's radiant beauty. The waters here are not always calm or kind, but their continual pounding rhythm sets the pace of life for the people of the region.

45 Bottom left and right
The sun-dappled hills of Ireland rise above the deep blue ocean waters at Sligo Bay, a favorite subject of Ireland's beloved poet, William Butler Yeats. Dotting the landscape of the Irish coast are beautiful, serene farmsteads, which are separated by fieldstone walls.

46–47
Farmers' fields border the ocean on St. John's Point, a seven-mile-long finger of land along Donegal's northern Irish coast. Just off the peninsula lies Inver Bay, with the larger Donegal Bay beyond. Sandy beaches are interspersed with the rocky shoreline, at some points rising to high cliffs. Most of the coastal villagers farm, fish, or cater to tourists.

46 Bottom left
The weathered rock formations of Skellig Rocks lie off Ireland's southwest coast and St. Finian's Bay in Kerry County. Seabirds nest in the rock crevices, which beautifully reflect the play of light and shadow during the course of the day.

46 Bottom right
A fine example of Ireland's imposing sea cliffs, the Cliffs of Moher tower above the ocean waters in County Clare along the central west coast. The cliffs look out over the Aran Islands.

47 Top
Ancient forts line the shores of stark Inishmore, the largest of the three Aran Islands. Located at the entrance to Galway Bay, the islands also include Inishmaan and Inishere. Robert Flaherty immortalized the fishermen of the islands in the 1920s documentary Man of Aran.

47 Bottom
The Irish coast ends in craggy peninsulas and offshore islands, many of which are uninhabited. This beautiful section of the coast is located on the Dingle Peninsula. The view includes Rough Point at the entrance to Tralee Bay. Heather, primroses, and bluebells crown the cliffs, while anglers try their luck on the beaches below.

THE WEST COAST OF IRELAND

heather, with peat beneath it. Sedges, rushes, ferns, and grasses grow in the turf. Eighty-one percent of Ireland is devoted to pasture and cropland, and the peninsulas of the west coast are no exception. From late spring to early autumn, visitors are understandably impressed by the magnificent wildflowers that grow in the hedgerows and fields leading down to the sea, including heather, primroses, bluebells, foxgloves, and fuchsia.

The landscape is probably best known to outsiders through motion pictures such as Ryan's Daughter and Far and Away and others that were filmed there. The emerald green fields are bordered by low stone walls, making the landscape look like a patchwork quilt, squared off in a grid pattern of green broken by dark gray walls. The fences run up to the coastline and then suddenly stop at the dramatic cliffs standing high above the ocean. Successive waves of peninsulas can be seen from many locations, seemingly stacked one upon another.

And then there is the ocean, continually crashing onto the shore, at times with unbelievable ferocity. The eroded brown canyons leading to the beaches open out on miles of soft sand backed by black rock, with cliffs towering above. Wave after wave crashes to shore, through storm and calm, bright summer days, a never-ending rhythm in this part of the world.

Many of the villages and lifestyles have changed little, and traditional Irish architecture, including buildings with thatched roofs, can still be found. Most residents speak the native Gaelic tongue as well as English.

There is much to do in western Ireland in addition to admiring the scenery. Walking, horseback and bicycle riding across the gorgeous landscape, ocean swimming, canoeing, angling, diving, and windsurfing are all popular. Enjoying these sports, exploring the wild natural beauty of Ireland's west coast, and learning about the area's fascinating history combine to make a rich and memorable experience.

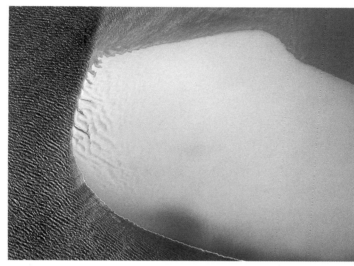

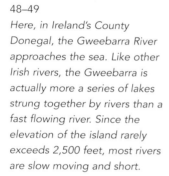

48–49

Here, in Ireland's County Donegal, the Gweebarra River approaches the sea. Like other Irish rivers, the Gweebarra is actually more a series of lakes strung together by rivers than a fast flowing river. Since the elevation of the island rarely exceeds 2,500 feet, most rivers are slow moving and short.

49

Three views of the Gweebarra River show its slow approach to the ocean waters on the west coast of Ireland. All the major rivers of Ireland flow from the interior plain. The largest, the Shannon and the Erne, follow this pattern. Despite its great beauty, Ireland has been the scene of many misfortunes. Most notable was the potato famine of 1849, which dramatically reduced the island's population through death and through emigration. The island remains largely rural today, particularly the west coast, and is a delight to the tourist, who may see it by auto or bicycle, or on foot.

ATLANTIC
OCEAN

NORTH
SEA

GERMANY

MEDITERRANEAN SEA

THE POSTCARD GERMANY

THE BLACK FOREST

Over several centuries, the Black Forest has played a huge role in Germany's culture and has been celebrated in everything from the country's famed romantic poetry to its fairy tales. It is little wonder that such frightening tales as "Hansel and Gretel" were set here. After all, it is called the Black Forest since so little light reaches its floor.

Located in the southwestern corner of Germany near the French and Swiss borders, the area was somewhat inaccessible until the turn of the twentieth century. Although it has many of its own traditional architecture and clothing styles, and even various regional accents, in many ways the Black Forest has

come to epitomize Germany and things German. Many first-time tourists, having seen photographs and paintings of the countryside and having read descriptions of its beauty, think of it when they think of Germany. Luckily they can easily explore the splendors of the Black Forest by hiking through it on many beautiful trails.

The Black Forest covers a mountainous region of granite and sandstone with deep, narrow valleys. The region averages 2,500 to 3,000 feet above sea level, with some peaks rising to 4,000 feet. The rocks that make up the mountains date back to the Devonian Period of about 408 to 360 million years ago. At that time an ancient mountain belt, the Hercynian, was pushed up by heavings within the earth. Remnants of the Hercynian Mountains exist today in the Armorican Massif, the Vosges Mountains, and the Black Forest. Technically the Black Forest rests on a rock formation called a horst, which is a horizontal block left standing between two valleys. (Korea and the Sinai Peninsula are two other locations with horst faults.) The region

is bordered by the Rhine River on the south and west. The sources of the Danube River rise on the east side of the forest's mountains.

There are many ski resorts and spas located in the Black Forest and although the height of the mountains cannot compare with the nearby Alps, they are nonetheless popular. A network of incredible hiking trails exists in the forest, making it possible for everyone from a serious backpacker to a casual hiker to enjoy the sublime scenery of wild mountain streams, interesting rock formations, and huge ancient trees.

The pleasant forests are cool and enjoyable in the summer, and damp and chilly in the snows of winter. The trails are covered with soft pine needles and are well maintained. The straight, stately trunks of the firs and pines contrast with the deciduous trees of many varieties. Gorgeous scenic views of rolling tree-covered hills and valleys recede into the distance, giving one the sense of being in an endless forest. Streams and waterfalls reflect the light of the sun onto the surrounding trees and rocks. The birds sing and chirp in the branches overhead, while small animals like squirrels and mice rustle in the underbrush.

In the winter snow crunches underfoot and the shushing of cross-country skis is heard. The forest sounds different in the winter. Noise seems to travel more quickly, and it seems more staccato. The evergreens look darker and they stand out starkly against the snow. The partly frozen creeks spill water over frozen ice and rocks, with the crystals of snowflakes gathering for a split second on the dry brown grasses.

One of the most scenic routes through the Black Forest, winter or summer, is the road from Waldkirch to Hinterzarten, from which the Alps can be seen to the south and the Rhine valley to the west. Another route, from Baden Baden in the north to Freundenstadt in the south, runs along the mountaintops overlooking the Rhine Valley and affords views of the distant Vosges Mountains in France. The isolated and remote villages in the southern Black Forest, also called the High Black Forest, are filled with rounded mountaintops and thick trees. This region is known as a *Wanderparadies*, or hiking paradise, and short hikes can be made through beautiful forests from one picturesque town to the next. Triburg, a town surrounded by steep hills, has Germany's largest waterfall, which drops 527 feet in seven stages.

The Black Forest is an area of great cultural importance and natural beauty and certainly what many visitors expect Germany to be. Enjoy it on foot—aside from the delightful exercise its wonderful trails provide, it will satisfy a yearning to be close to nature.

50
Autumn breezes rustle through the changing leaves on trees in the beautiful meadows and farm fields near St. Peter in the High Black Forest. This is the region called a Wanderparadies, or hiking paradise, by the Germans.

50–51
Wet snow clings to the trees and meadows of Kandel Mountain in Germany's Black Forest. Located in the southwestern corner of the country, the Black Forest provides postcard views and dramatic surprises to all of its visitors.

51 Bottom left
A typical Black Forest scene made up of mixed, old-growth forest and open meadowland teeming with wildflowers spotlights this spectacular region of Germany. This scene is near Baiersbronn in the northern Black Forest.

51 Bottom right
An Alpine wonderland spreads before the hiker in the High Black Forest, the sun warming the face while the brisk breezes turn cheeks a ruddy red. This panoramic view was taken from Belchen Mountain in the Alps.

CENTERPIECE OF EUROPE

THE ALPS

Perched high upon a mountain ridge, the deep, light green valleys stretch out before you, tiny villages looking like a cluster of dollhouses scattered here and there over the face of the land. Dark green, spruce-covered mountains rise up behind, topped by sheer, gorgeous blue peaks covered with snow. Clouds hover close above the mountains, white vapor mixing with the pristine ivory of the snow caps. The peaks recede into the distance, colors becoming more pastel and muted with distance. Waterfalls appear from beneath dark trees on the mountainside, trailing downward in ribbons of white to the rivers below. A bell clangs as a curious cow approaches, while the wind blows

lightly on a warm summer day. The ridge is covered with wildflowers, including the deep red alpenrose and the lovely multicolored gentians. These are the magnificent Alps, the mighty mountain chain that towers over all of Europe.

The Alps constitute one of the world's best-known and best-loved groups of mountains. Important historically, culturally, and visually, the Alps have had a profound effect on world history, dividing the very center of Europe. They extend in a 750-mile arc from Vienna in the east to Marseilles in the west, and lie within France, Germany, Austria, Liechtenstein, Italy, Slovenia, and Switzerland. Covering a total of 92,700 square miles, they are the highest mountains in Europe and with twenty million people probably the most densely settled in the world. Dairy farming, hydroelectric power, and the mining of salt and iron ore support the native economy, with a huge assist from tourism. People from around the world come to see the Alps, which for many visually define what a mountain range is.

The Alps were formed when the continents of Africa and

52–53 Top
This dramatic panoramic view taken in the Bernese Oberland of the mountain known as the
Finsteraarhorn provides a sample of the region's incredible scenic and natural wonders.

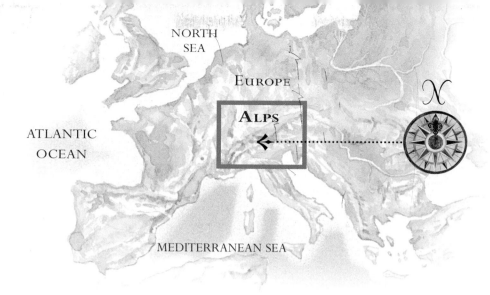

52 Bottom left
An awesome landscape of rock, snow, and tress can be seen from Kandersteg, a small town in south central Switzerland in the heart of the Bernese Alps. Looking upward, the Doldenhorn towers over the town at an altitude of more than 11,900 feet.

52 Bottom right
On the border of Bavaria and Austria is Zugspitze, Germany's highest mountain at 9,721 feet. Skiing resort complexes like the one in the foreground bring people here above the clouds for rest and relaxation, as well as for sport. In the summer, dramatic hiking trails canvass the area. Zugspitze is just above Garmisch-Partenkirchen and less than fifty miles southwest of Munich.

52–53 Bottom
Mountain climbers reach the summit of one of the jagged, towering peaks of the Mont Blanc massif, Aiguille Maudit, mantled in perpetual snowy whiteness. The town of Chamonix below Mont Blanc is often crowded with those who want to test their mettle on the mountain. Englishmen, seeking the excitement of the sport, began climbing the peaks in the 1850s.

54 Top left
The delicate reflections of light off the snowy sides of the north face of Monte Rosa stand out sharply against the brilliant blue sky, which highlights the mountain's distinctive shape. The profusion of such amazing mountain peaks dazzles anyone looking at Europe's spine.

54 Top right
This spectacular view shows Monte Rosa in all its glory, dramatically lit by the fading sunlight, backlit with the mysterious glow and banded clouds of the end of the day, its rough texture tempered by the smooth drifts of snow atop the glaciers. The Alps are beautiful, but they can be harsh, especially to visitors who arrive unprepared for their rigorous conditions.

54 Bottom
The distinctive pyramid of the Matterhorn, towering 14,690 feet above sea level, dominates Italy's Aosta Valley. The incomparable beauty of the jagged mountain peaks dividing the light from the darkness has inspired artists for hundreds of years. Valle d'Aosta lies below the Pennine Alps, including Mont Blanc on the west and the Matterhorn and Monte Rosa on the east. The Graian Alps run through the center of the province.

Europe began colliding about forty-five million years ago. The oceanic crust between the Adriatic Plate and Europe's rim was folded up, causing the uplift of the Alps. Folding occurred along the edges of the colliding continental plates, deep underground, turning a shallow sea into mountains. The African continent continued to move at the rate of approximately two inches per year, pushing into Europe a distance of over 150 miles and piling up the Alps. Over millions of years, pressure and high temperature caused rocks that were normally brittle to bend instead of break. This process produced the repeated folding of the Alps. The folds were sheared through and a side of the fold was forced upward as much as several miles. This type of fold is called a nappe. The stresses of the earth during the creation of mountains like the Alps produced the metamorphic rock that is found so abundantly there.

As the earth's history goes, the Alps are fairly new mountains, and they continued to grow and change for more than twenty million years. A single last uplift phase about three million years ago made the Alps one of the world's highest mountain ranges. The final phase in the geological drama of the formation of the Alps was glaciation. Four separate glacial events sculpted valleys and left deposits behind, which dammed rivers. This created the region's many lakes, including Lake Geneva and the Lake of Constance, or Bodensee.

While hiking or scaling the Alps, one passes through several climatic regions, each with its own natural beauty. At the base, forests of oak, linden, maple, and pine alternate with verdant farm fields, orchards, and vineyards. Reaching the altitude of about 4,900 feet, the scene changes as the broadleaf trees disappear and a coniferous forest takes their

54–55
Europe's second highest mountain, the Monte Rosa massif, catches the rosy glow of sunset. Monte Rosa is Switzerland's highest peak, at 15,203 feet. The massif contains

several individual peaks named Parrotspitze, Punta Gnifetti, Zum Steinspitze, Dufourspitze, and Norden. Europe's longest ice flow, the Great Aletsch Glacier, runs to the Rhone Valley below.

55 Bottom
Looking down the length of the Monte Rosa massif, mountain after mountain can be seen as the Alps fade away in

the distance. In the center foreground is the Capanna Margherita hiking shelter, the highest in Europe at more than fifteen thousand feet.

place, composed of larch, cembra pine, and Norway spruce. This area is not all forested, and farm fields, hay meadows, and pastures exist up to the tree line at 6,000 to 6,900 feet. Above the tree line there are gnarled, stunted trees and dwarf shrubs, then open alpine pastureland and grasses. It is at this level that the climber receives a panoramic, unobstructed, and wondrous view. Mountain flora have adapted to extreme climatic conditions similar to tundra vegetation. Colorful wildflowers such as alpine columbine, ladies slipper, spring anemone, edelweiss, and trumpet gentian delight the climber in spring and summer. One might also catch a glimpse of some of the animals of the region browsing for food. Most likely these will be domestic farm animals like cows, but a lucky climber might spot an ibex, chamois, Alpine marmot, ptarmigan, wood mouse, or Alpine chough. Climbing even higher, the visitor notes that between 8,200 and 9,800 feet all

56–57

The Freney side of Mont Blanc beckons mountain climbers, daring them to conquer her slopes. But as history has shown, even the most experienced climbers must beware. The weather can change suddenly, and snowstorms may occur at any time of the year, turning an exhilarating climb into a life-or-death situation.

58 Bottom left
The drama of the Dolomites includes areas of smoothly sculpted snow that look more like a fabulous dessert than a gigantic, immovable work of nature. From the top of such a peak, a placid mosaic of sun-dappled, green valleys and farm fields can be seen.

58–59
The haunting beauty of the Dolomites can be seen in this view of the rugged region. The Dolomites are a chain of mountains that form part of the eastern Alps in Italy. The Brenta peaks divide four valleys in the Trentino region of northeastern Italy, due east of Trent.

59 Bottom left
The incredible Dolomites form a long chain of mountains that point to the northeast, toward the Austrian Tyrol. Water collects in the cracks of the mountain rocks, then freezes, eventually splitting boulders apart. Climbers must wonder at these incredible natural forces.

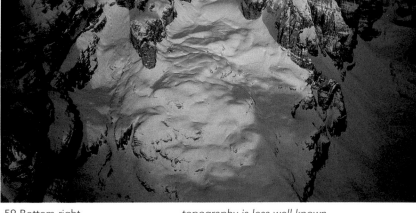

59 Bottom right
Each morning and evening the Dolomites are shrouded in fantastic colors as they reflect the growing and waning light of day. Their incredible topography is less well known than that of the Swiss, Austrian, and German Alps, primarily due to the fact that they are not as easy to reach via public transportation.

60–61
The Catinaccio Group of mountains, specifically the Torri del Vajolet, stands 9,856 feet above sea level. These mountains are located about ten miles due east of Bolzano, the central city of the Dolomite range.

60 Bottom left
The wild and uninhabited reaches of the Pale di San Martino can be seen from the town of San Martino di Castrozza. Evidence in the rocks reveals the geological history of the region, including the crustal deformations that continue to this day.

60 Bottom right
The Piz Boe group of mountains in the Dolomites as seen from Pordoi Pass sparkle like a scene from a fairy tale. Each portion of the Dolomite range has its own unique character, and every sector is captivating because of its great beauty.

61 Top left
The face of Mount Pelmo in the Cadore area of the Dolomites changes constantly as the shifting sun alters its colors.

61 Top right
In the northeastern part of the Alto Adige lies the Badia Valley, with the Sasso Della Croce rising above it. This scene is located at the edge of the Dolomites along the Badia River, which runs south to north.

white when snow covers the ridges in the winter. The Austrian peaks have an extensive network of hiking trails complete with mountain refuges called Hutte. Alpine associations in Austria maintain over eleven thousand of these Hutte, which provide overnight accommodations, cooking facilities, and in some locations hot meals. The area is also a skier's paradise, particularly around Innsbruck and Kitzbuhel in the Tyrol. Grossglockner, a peak in southern Austria, is that nation's highest point and includes the massive Pasterze Glacier.

The Dolomites, located east of the Adige River in Italy,

61 Bottom
From the town of San Martino di Castrozza on the eastern side of the Dolomites, one looks upward toward the Pale di San Martino.

At dawn and dusk the Dolomites appear to be lit from within as the light makes their limestone change to hues of purple, red, orange, and pink.

present a unique and fascinating sight, with their pinnacles of limestone. In some areas the amazing peaks seem to shoot skyward from pine forests and farm fields. The white rock of these peaks is accented by the green, sunny, rolling hills beneath. Some of the peaks are marked with dolomite, a type of limestone, which provides them with unusual colors that change with the rising or fading light of day. When the sun glints off the rock at midday, the light may be almost blinding. Yet, as the sun sets its rays are filtered through the atmosphere, and the color of the rock seems to change from yellow to orange to deep red. The peaks themselves are wonderful rock fantasies, some looking like church spires and others like pyramids.

The mountains and the type of rock were named after the French geologist Gratet de Dolomieu. The tallest peak in the Dolomites is Mount Marmolada, 10,965 feet above sea level. The sheltered valleys that open from these peaks onto the upper Italian lakes showcase subtropical vegetation in contrast to the snow-covered mountains in the distance.

From any point of view, the Alps create a beautiful backdrop, an incredible centerpiece for the continent of Europe. They are not only one of the world's most beautiful areas, but also one of the most culturally important. The Alps provide beauty and drama for the millions of resident Europeans and visitors from around the world who enjoy them annually.

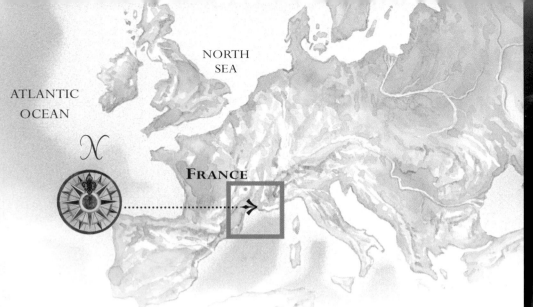

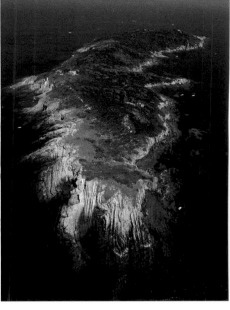

SUN-DRENCHED
REGION of COLOR

FRANCE'S MEDITERRANEAN COAST

Seabirds wheel and call overhead, while the rhythmic pounding of waves lulls sunbathers to sleep. The smell of salt water is in the air, the beaches are fine, the sun is warm, and the flowers at the edge of the sand are fragrant. The waves and the cloudless sky seem to meld perfectly, from the white of the breakers, to the greenish-blue of the waters, to the deep blue of the sky. The Mediterranean coast of France, stretching in a long crescent from Italy on the east to Spain on the west, is beautiful year round.

Part of this gorgeous coast is known as the Riviera, a narrow coastal strip running along the Mediterranean Sea from Hyères, France, to La Spezia, Italy. The Alps shelter the

Riviera from cold winds during the winter months, providing a mild year-round climate that allows tropical fruits like pomegranates, bananas, and dates to flourish, as well as lovely palm trees. An inordinately large number of sunny days provides good growing conditions and excellent light. Artists like Van Gogh and Cezanne agreed, and flocked to the region to paint it. The ocean views are superb, the beaches sandy, the temperature just right for vacationing and relaxing. But to truly appreciate the Riviera one must go beyond the tourist spots. High above the water, coastal towns cling to the edge of dramatic cliffs. From the surrounding hills, breathtaking views of land and water are available. Hiking above the shores provides a view of the real beauty and color of the region.

Offshore from the French city of Hyères lie three islands that present an image of an older Mediterranean, with sparse populations and dense forests. These islands have been nicknamed The Golden Islands by the French, perhaps because of the golden glint of mica schist in their southern cliffs. It is here that the natural aspects of the French coast can be truly appreciated, which may also account for the nickname. Although the smallest of the islands, Ile du Levant, is primarily a military reserve, the wild landscapes of all three have been preserved as natural parks since the 1960s. The largest of the islands is Porquerolles, just four miles long and two miles wide. Its gentle sandy beaches contrast with thick, fragrant inland vegetation of pine, eucalyptus, lavender, heather, and rosemary. Migrating birds stop over each spring, and the waters teem with the creatures of the ocean, protected from spearfishing by law. Port Cros, the second largest island, is a beautiful, hilly nature reserve, the smallest national park in France. Footpaths plunge into its forests and around its bays, and an underwater nature trail allows snorkelers to explore its marine life. The aptly-named Vallon de la Solitude lies at the center of the island.

Moving west along the Mediterranean coast toward Marseilles there is an incomparable view from the wall of towering sea cliffs called Cap Canaille. At 1,300 feet, it is Europe's highest cliff. It provides a dreamy overview of the mazelike coastline of Les Calanques and its offshore islands, which seem to float above the ocean waters. Les Calanques,

62 Top
Port Cros is the second largest of the three Porquerolles Islands. It is set aside entirely as a nature preserve and is the smallest National Park in France.

62 Bottom
Capo Rosso is located on the craggy west shore of Corsica in the central portion of the island near the town of Piata.

62–63
The Esterel highlands along the French Côte d'Azur rise between Frejus and St. Raphael on the west and Cannes on the east. Views are breathtaking from the road as motorists drive between the sea and the red rocks. Hikers are treated to an even more spectacular treat from such vantage points as Pic du Cap Roux, seen here.

63 Bottom right
The Esterel highlands provide scenic views of the ocean as well as the land. The sweetly scented vegetation and brilliant red rocks contrast with the deep blue of the ocean, providing an unforgettable visual treat.

63 Bottom left
Porquerolles Island off the French coast teems with vegetation, including lavender and other native plants. Porquerolles, four miles long and two miles wide, is the largest of the three islands dubbed Les Iles d'Or, or The Golden Islands.

64–65
The flat and swampy lands of the Camargue support a profusion of migrating birds, particularly waterfowl. The area's lovely shallow waters reflect the light of day in extraordinary ways. Sunsets are particularly gorgeous in this marshland by the sea.

64 Bottom left
Each year, thousands of elegant pink flamingoes migrate from Africa to the marshy Camargue on the southern coast of France. The area is a nature preserve and contains farmland. It is also supports some industry, such as extracting salt from the ocean.

64 Bottom right
The Camargue is a pocket of land that differs in many ways from the surrounding areas. In many ways it is like another country. It is a land where black bulls are raised alongside unique white horses that are herded by gardians, the French cowboys.

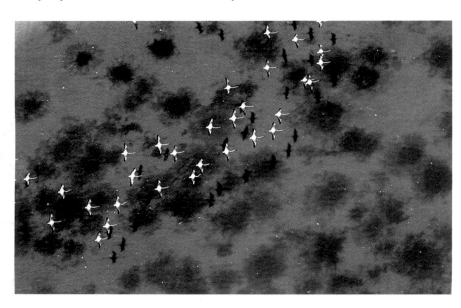

65 Top left
Along the French coast near the city of Marseilles are the Calanques, rocky volcanic outcroppings framing picturesque inlets, many accessible only by boat. As waves crash onto secluded, sandy beaches, their sound echoes from the surrounding cliffs, creating a wonderful sense of solitude and peace .

65 Top right
The Calanques are located between Marseilles and Cassis, and are serene and intriguing places to explore, whether by paths from above or from boats on the water. Riding slowly in a boat as it gently rocks in the warm blue water makes for a restful afternoon.

65 Bottom
This long strand of beach fronting the Lyon Gulf is the Étang de Thau, located between the town of Sete and the Cap d'Agde and west of the Camargue. Inside the bay the water is fed by tributaries of the Herault River. On the outside, the Lyon Gulf is situated on a very shallow portion of the continental shelf, where ocean depths rarely exceed fifty feet for fifty miles out to sea.

FRANCE'S MEDITERRANEAN COAST

with their volcanic rock walls, are rocky inlets inaccessible from the roadless coastline above, and their beaches can only be explored by boat. These beautiful areas of sun-drenched rock and pounding surf form little havens of paradise for those who love the scents, flora, and fauna of the sea. Les Calanques were formed when the sea invaded narrow river valleys between these cliffs at the end of the last Ice Age. The dramatic white cliffs and needlelike rocks rise dramatically from the sea and can be investigated on footpaths running along the top. Nature bestowed an incredible gift on this area in its incomparable weather and in the the geological conditions that created it.

West of Marseilles lies the wide circular bay of the Camargue and the flat lands surrounding it, embraced by the delta of the River Rhone. Farmers created the lands of the Camargue when they drained the delta wetlands centuries ago and turned them into arable crop and grazing land. Today, a 205,000-acre French Regional Park protects much of this region. The low-lying lagoons and lakes are home to flamingos, egrets, and herons, and are a favorite of bird watchers. But the Camargue is probably most famous for its black bulls and pure white horses, often ridden by gardians, or French cowboys. The region is home to a wealth of wildlife, including wild boars, badgers, tree frogs, water snakes, and pond turtles. Juniper trees up to seventeen feet tall border the fields and seem to divide them from the sea. The central islands surrounded by the marshland of the Camargue are completely protected and out of bounds to humans.

Sun, sand, and surf blend together to produce a world of sunlight and laughter along France's breathtaking Mediterranean coast. Spectacular mountains and cliffs rise high above the azure ocean waters, with villages perched precariously above. These elements combine magically to make this a glorious and unforgettable coastal paradise.

66–67
Many portions of Corsica resemble island paradises in the South Pacific or the Caribbean. The calm waters and warm temperatures of Seleccia Beach in the Desert of Agriates, are perfect for swimming, boating, and other water sports.

66 Bottom left
The waters of St. Jolie Bay on the west coast of Corsica seem to glow in the sunlight as people bask on the beaches and sailboats.

66 Bottom right
Bonifacio, Corsica, is located on the southern tip of the island, just nine miles from Sardinia. Corsica is the fourth largest island in the Mediterranean, and features azure mountains, green rolling hills, bold red sea cliffs, and white sandy beaches.

ATLANTIC OCEAN

FRANCE

ITALY

CORSICA

MEDITERRANEAN SEA

N

THE MOST BEAUTIFUL ISLAND

CORSICA

The high, dry white cliffs tower above you as you plunge into the warm ocean waters below. Heather and prickly pear head the sandy beach, while the sedimentary layers of the cliffs beyond look like a layer cake. The breeze slowly rustles the leaves of scrub oak trees as sailboats tack to fill their sails with wind on the distant blue water. Land of old fortifications, splendid rock formations, colorful fishing villages, and vacation spots, backed by verdant vineyards on the hills above, this warm oceanic gem is Corsica.

The Greeks called it Kalliste, or most beautiful. The largest of France's offshore islands, Corsica, or Corse as it is called in French, includes 3,350 square miles of varied and beautiful topography. The island is located south of France and parallel to the western Italian coast. It is separated from the Italian island of Sardinia to the south by the narrow, nine-mile Strait of Bonifacio. Jagged purple mountains loom above its coastal hills, with towns and homes perched above crystal-clear lagoons and bays bounded by spicy scented maquis, or heavy undergrowth. The west coast of the island is rocky and indented with bays and harbors, while the east coast tapers down to the low coastal plain of Aleria with its lagoons and swamps full of graceful water birds. Short, powerful streams rush from the central mountains covered with oak, chestnut, and cork trees. The Golo and Tavignano rivers are the island's largest. From Corsica's tallest point, 8,892-foot Mount Cinto, the full sweep of this beautiful island, birthplace of Napoleon, may be taken in with a turn of the head. The island's interior is wild and rocky, marked by narrow, fertile valleys alive with scrub. The mild climate and rich soil allow a number of crops to be grown, including grapes.

But it is the 275-mile-long coastline with its beautiful turquoise waters that lures tourists and vacationers. Although most of the coast is high and craggy, with few natural harbors, plenty of beach and recreational areas exist for swimming, boating, and other water sports. Colorful villages, including many isolated fishing ports, are home to people who make their living not only from lobster and anchovy fishing, but also from tourism.

The eastern shore includes the beautiful "Costa Verde" or green coast. On a clear day the Italian mainland is visible from here. Just inland is the beautiful Cascade of the Buccatoghju, one of many lovely waterfalls on the island. As their leaves rustle in the wind, eucalyptus trees scent the air dazzling hikers.

The northwestern coast near Calvi is warm with the aromatic scents of scrub and seaweed. The gentle hills of the Balagne Valley are covered with olive, lemon, orange, and fig trees and

67 Top
High red rock faces and gorgeous secluded coves highlight Scandola National Park on the island of Corsica. The island's pleasant climate, rugged scenery and colorful villages draw thousands of visitors each year.

67 Bottom
This aerial photo of the west coast of Corsica dramatically reveals the shallow waters and rocky shoreline that allow the island few natural harbors. Nevertheless, Corsica is a favorite spot for pleasure boating, particularly sailing.

69

The rocky islands and peninsulas at Lavezzi on the Corsican coast invite sunbathers and *snorkelers. Their barren appearance is more than compensated by the thick vegetation of the mountainous interior.*

68 Top left
The rocky shoreline of Lavezzi Island is lapped by the waves of the Mediterranean Sea. Lavezzi presents a contrast to the east side of Corsica, which is marshy and low to the ocean.

68 Top right
Bonifacio is perched on a limestone cliff high above warm Mediterranean waters. The town overlooks the Strait of Bonifacio.

68 Center
The Golfo de Santa Giulia on the west coast of Corsica highlights *the sweeping, undulating nature of the island's 275-mile coastline. Inland, the wild, rocky highlands are covered with scrub and slashed with narrow, fertile valleys.*

68 Bottom
Complex layers of sedimentary deposits score the cliffs near Bonifacio. They reveal its history as a site associated with the ocean for millions of years. Topped by scrub growth known as maquis, the island includes 3,350 square miles of varied and beautiful scenes like this one.

CORSICA

are often referred to as the Garden of Corsica. The spectacular coastline also includes the red granite rock formations of the Calanches of Piana, said by Guy de Maupassant to look like the "fantastic people of a fairy tale, petrified by a supernatural power." Today these wonders lie within Scandola National Park, and include wildlife, caves, and the rocky peninsula itself. Created in 1975, Scandola is part of a network of national natural parks that preserve Corsica's complex environment. Throughout the island, fallow deer, boars, foxes, weasels, and wildcats are protected to boost their numbers. Bird life is also protected, particularly the huge lammergeier, with a wingspan of nine feet, and rare breeding pairs of golden eagles, white-tailed eagles, and ospreys, which make Corsica their home. These parks preserve marine life as well. A profusion of beautiful and varied seaweed, coral, and fish species live and breed offshore, many of which, like the grouper, dentex, and corb, have become rare in the Mediterranean.

To the south, the city of Bonifacio is exposed to the open sea and fierce winds that whistle through the strait. It seems perched precariously on the sedimentary cliffs. Donkeys once brought necessary goods to the upper town by means of a winding road, which is still in use today.

South of Bonifacio are the Isles of Lavezzi and Cavallo, set amid dangerous reefs and rocks that have spelled the end of many ships. The islands themselves are but masses of rock jutting out of the sea. They provide a perfect platform for water sports and anchorages for sailboats enjoying the beautiful Mediterranean weather. The deep turquoise of the warm ocean calls to the visitor, with fish and other sea creatures easily visible below the water's surface.

This, then, is Kalliste, thought by the ancients to be the most beautiful of islands, noted for its greenery in the midst of the ochre Mediterranean lands. Corsica delights the eye and the imagination of visitors with a perfect combination of naturally beautiful sights, scents, and sounds. It is an island experience par excellence, with the gorgeous vistas of nature's wonders surrounding all who visit.

70–71
A little group of islands off the northern coast of Sardinia, the Maddalena archipelago is noted for its crystal clear waters and rocky shores. One of the islands, Caprera, was the birthplace and final home of the Italian hero Giuseppe Garibaldi, the flamboyant general who helped unify his nation in the 1860s.

70 Bottom left
The typical low vegetation of the Sardinian coasts can be seen in this view of the Island of Budelli. Fragrant plants that grow on the hillsides are called macchia.

70 Bottom right
This view of the shallow ocean waters of northern Sardinia shows the narrow straits between the rocks and small islands of the Maddalena archipelago. The smell of the salt air and the fragrant greenery mixes with the sounds of the waves slapping against the shore to create a memorable experience in this aquatic paradise.

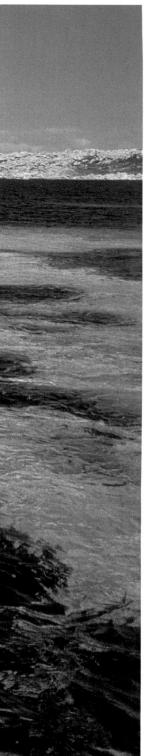

ATLANTIC
OCEAN

FRANCE

ITALY

SARDINIA ←·····················

MEDITERRANEAN SEA

A CONTESTED JEWEL

S A R D I N I A

It is difficult to describe the waters of the Mediterranean off Sardinia. They are too clear, too turquoise, too pristine. The sunshine and the white sand are too perfect. Walking on the beach or swimming in these waters seems more like a fantasy than reality. Yet Sardinia is a real place, a large island surrounded by a gorgeous coastline that reveals greater and greater beauties around each bend.

Sardinia is a complex and many-faceted island off the west coast of Italy, nine miles south of the island of Corsica. It is the second largest island in the Mediterranean Sea, 166 miles long and 75 miles wide, with an area of 9,301 square miles. Sardinia has been contested over and over again by various nations and powers during its long history. The struggle for possession began with the Phoenicians, and continued with the Carthaginians, Romans, Vandals, Byzantines, Saracens, Pisans, Genoese, Spanish, Austrians, and finally the Italians. It is no wonder they were all so anxious to have it.

The island has over one thousand miles of shoreline, with pristine, white sandy beaches running for mile after mile along its coasts. The azure blue waters are clear and warm, supporting all kinds of water sports. Most tourists are drawn to the eastern coast of the island, although many people try to explore less crowded areas in different corners of Sardinia. The brilliant blue skies, clear waters, and white sand beaches trailing off into the sea create an ocean paradise difficult to resist.

Many coastal areas of the island have escaped urbanization and feature nearly deserted stretches of beach that run for miles, interrupted only by occasional cliffs. Southern Sardinia includes beaches, hills, and inland salt lagoons with huge numbers of birds, including flamingoes. The Costa Verde on the southwestern corner of the island features slopes covered with arbutus and lentisk, providing the greenery that led to the area's name. These bushes lead to the beach and protect herds of deer and other wildlife. The area is full of artifacts recalling its use in mining for lead and zinc for thousands of years, from the time of the Romans until about 1960, when the mines closed.

A particularly spectacular stretch of coast begins at Piscinas and runs for over five miles to Scivu. This incredible beach area includes three thousand acres of golden sand

71 Top
Flowers grow along the sand dunes in the northern areas of Sardinia, nicknamed the Emerald Coast. This incredibly beautiful region is extremely popular with affluent tourists.

71 Bottom
The pink beach of the Island of Budelli, in the Madaalena archipelago, was subject to vandalism until the Italian government recently made it a protected natural area. Because the pink sand was so unusual, tourists would scoop it up by the bagful and take it away.

72 Top
These rocks are in the Nebida Gulf, which shows a rockier and more forbidding aspect of the Sardinian Coast. Unlike other other parts of the island, this region does not have sandy beaches.

72 Bottom
The Maddalena archipelago consists of rocky islands that can be explored on all-day boat tours run by local operators.

72–73
A close view of the famous pink beach on the Island of Budelli shows that the sand itself is actually pink. In the shallows, even the water reflects the pink.

SARDINIA

dunes and has been nicknamed the Sahara of Italy. The Mistral winds shape these hillocks of sand and permanently bend sweetly scented junipers and other trees. The smell of the salt air blends with the sand, the vegetation, and the sunshine to create a comfortable, wonderful sensation.

Although most tourists never leave the beaches, nine-tenths of Sardinia is actually composed of mountainous terrain. The highest point on the island is Monte del Gennargentu, towering over the adjacent countryside at 6,017 feet. The mountains can be difficult since the slopes are steep and heavy rainfall can lead to heavy erosion and even landslides, which have created the picturesque undulations of the territory. Despite these occasional dangers, Sardinians continue to farm and cultivate the rough hills. The Campidano Plain on the southwestern part of the island is the principal farming area, where livestock, grain, tobacco, olives, and grapes are raised. Fishing and

mining for zinc, copper, and salt are also important industries. This is another Sardinia, where the topography and the lifestyle differ from those of the seacoast areas.

Sardinia also has vast wilderness areas such as Sulcis-Iglesiente and Supramonte-Barbagia-Gennargentu, noted for harsh, mountainous terrain and the incredible play of light and shadow on its hills. Wild ponies roam the Giara di Gesturi, a rough, rocky plateau area.

The Sardinian writer Marcello Serra entitled one of his books *Sardinia: Almost a Continent*. He explained that Sardinia is unique and possesses so much cultural and natural variety that it is similar to most continents. Sardinia is indeed an extraordinarily beautiful island with clear bright sunshine and incredible vistas, whether they be in the lofty hills and peaks or along the sandy beaches on the sea.

73 Bottom left
The landscape around Lula, with Mount Turuddo and Mount Albo in the background, indicates the great variety of topography and vegetation that can be found on the island of Sardinia.

73 Bottom right
Dune di Piscinas are beautiful sand dunes that characterize southern Sardinia in the Iglesiente area. The beautiful green-covered hills extend back to distant purple peaks, which are in the interior of the island.

74-75
A maze of trees and shrubs compose one of the highlights of the Uccellina National Park on the Tuscan coast. This wild,

unspoiled region is carved from the Maremma, a region of salt marshes and canals that is home to wild boars and water birds. The park was created in 1975.

74 Bottom left
One of the great tourist destinations of southern Tuscany is Monte Argentario, a peninsula jutting from the Italian coastline with steep cliffs rising majestically above the ocean waves.

74 Bottom right
The red cliffs of the Island of Capraia, called Cala Rossa in Italian, are composed of dark granite. Although most of the island is a prison, its non-restricted areas are popular with vacationers.

75 Top
Beneath the waters off the Tuscan coast lies an incredibly rich and diverse world of ocean life. Offshore islands like Giannutri beckon tourists and recreationists to visit and enjoy their many wonders.

75 Center
The Castello di San Giorgio stands guard over the ocean cliffs on the island of Capraia off the Tuscan coast. The hilly island is punctuated by deep folds of earth called radi, covered with Mediterranean brush.

75 Bottom
The lovely island of Elba basks in the sun, covered by deep green vegetation. Mountains rise above the turquoise waters, while the island is ringed by

pleasant beaches. The stunning land and seascapes complement an island that was once a giant peninsula linking Corsica to the Tuscan mainland.

ITALY

TUSCANY

MEDITERRANEAN SEA

ITALY'S CULTURAL GEM

TUSCANY

tately cypress trees line the green fields of romantic Tuscany, a province in west central Italy. The setting sun makes the trees cast long, dark shadows across the fields, forming a herringbone pattern on the grass. Ochre-colored villas are spread out across the rolling hills, suggesting the area's colorful history. Tuscany is an area of Italy in many ways unchanged by progress. It continues to offer insight on the traditional and historic landscape of this vital nation.

Tuscany is rich both in culture and landscape. Visitors from around the world come to see Florence and its artistic treasures, Pisa and its famous leaning tower, and other famed cities and sites. But Tuscany is much more than this. Northern Tuscany has mountains and beaches along the Mediterranean Sea. Eastern Tuscany has the lush forests of the Mugello, while southern Tuscany, rarely visited by tourists, features interesting flora and unspoiled beaches.

Whether looking at natural features or those made by man, Tuscany offers idyllic views celebrated by poets and artists for hundreds of years. The daylight in Tuscany is like that nowhere else on earth. It is golden and sometimes hazy, giving the landscape a nostalgic feeling. This is true both in the countryside, looking at spacious farms with their olive groves and vineyards, or in Florence, looking at the historic buildings and churches from across the Arno. Tuscany is rich in wildlife and flowers, and the songs of cicadas and grasshoppers punctuate the sound of the countryside.

The northern area of Tuscany includes the Lucchese plain, and further north, foothills covered in olive groves that produce some of the finest oil in Italy. Above these hills are the wild and mountainous areas of Garafagnana, the Apuan Alps and the Lunigiana, rising over 6,000 feet. Castles, monasteries and villages dot the landscape amid the splendid scenery of white and pink mountain peaks and the deep green of the chestnut forests. Villages are located on pristine lakes, backed by blue mountains rising dramatically from the rolling hills.

The marble quarries of Cararra in the Apuan Alps have been in use since the time of the ancient Romans. The more than three hundred quarries make up the oldest industrial site in continuous use in the world. Artists have long carved the fine white marble of the area into incredible statues. The peaks of the Apuan Mountains rise suddenly from the rolling foothills, standing bare and gray above the trees. The white marble is embedded in the sides of the brown and gray rocks.

Northwest of Castelnuovo di Garfagnana are the area's highlands, preserved within the Parco Naturale delle Alpi Apuane, which was designated a nature preserve in 1985. The park encompasses deeply cleft valleys covered with trees, which lead to towering white cliffs that change to pink and violet with the shifting light of day. At 6,320 feet, Monte Pisanino is the highest peak in the area, towering above the placid blue Lago di Vagli. At Arni one can see the Marmitte di Giganti, or the Giants' Cooking Pots, great hollows left by glaciers of the last Ice Age.

The Versilia Coast includes a series of beach resorts on the startlingly blue sea, backed by green pine woods. The Apuan Alps form a dramatic backdrop to the scene of sandy beaches and large cliffs covered with green vegetation. The coastline is varied, with breathtaking scenes such as the old fort on the promontory of Monte Argentario that overlooks the circular

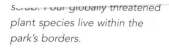
scrub. Four globally threatened plant species live within the park's borders.

ATLANTIC OCEAN

EUROPE

78 Bottom
Rolling sand dunes covered in green trees and brown grasses edge toward the ocean at Doñana National Park. The soil has a generally high...

ASIA

EUROPE

BLACK SEA

GREECE

ENDLESS AND SUBLIME VARIETY

THE GREEK ISLANDS

80–81
The famous Shipwreck Beach on the island of Zakinthos (Zante) is a spectacular landmark much admired by tourists and indicative of the incredible sights found in the Greek isles. Zakinthos is one of the Ionian Islands, located to the west of the mainland in the Ionian Sea.

81 Top left
Mirtos Beach is a narrow spit of sand on the large green island of Keffalonia. The island's wonderful beaches are complemented by a peaceful combination of sea and mountain scenery. The scent of pine trees fills the air, while eagles may be spotted soaring above. Keffalonia is located north of Zakinthos in the Ionian Islands.

81 Top right
The coast of the island of Corfu is seen here at Paleokastritsa, near the Grotta Bar. Corfu is the northernmost large Greek Island, mostly lying off the coast of Albania near the Strait of Otranto. Inland are fertile valleys with groves of olive and citrus trees and a wetland nature reserve.

81 Bottom
The smaller island of Atsitsa can be seen from Skiros in the northern Sporades. The heavenly scents of fragrant scrub line the hills above the beaches of the Greek Islands, which are famed for their incredible scenery as well as their recreational possibilities.

Moonlight shines through a building's latticework, through which a lengthy stretch of white beach can be seen. Mediterranean waves crash to shore, while steep cliffs jut upward, with little white houses clinging to their sides. Colorful fishing boats have been pulled up on shore, while in the distance hilly fingers of the land protrude into the water and crowd in upon each other. The silhouette of the cross on a church's dome can be seen, black against the azure sky. The night is warm and still, save for the rhythmic lapping of the water on the beach. These are but a few of the impressions one has while standing between town

and beach on a Greek island in the summer. And this incredible island is but one of thousands.

These special places conjure up images of tiny fishing villages with guest houses and taverns backed by the deep blue ocean and hills green with ancient olive trees. The scent of pine fills the air while one sits at a café in a shady little town square, surrounded by flower-bedecked trees. Creeks punctuate the coastline while eagles soar above. Waves crash onto perfect white beaches stretching off into the rosy sunset. Slim cypress trees border the fields, while an old fortress dominates the hillside. This is the land of philosophers, myths, hero tales, and gods. This is a land of sublime beauty.

There are over two thousand of these small, volcanic islands located between Europe and Asia in the Aegean Sea and along the coast of Turkey. Most are inhabited, although some are deserted. Most cater to the tourist trade, although

82
Located in the Aegean Sea, just
off the Thessalonian Coast of
mainland Greece, Skiathos is an
island in the Northern Sporades.

Skiathos includes fantastic
beaches, among the best in
Greece, as well as plenty of
forests and wildlife in the hills
above them.

the old occupations of fishing, sheep herding, growing olive
trees, and sponge fishing continue. The islands are grouped
into clusters, including the Sporades, the Cyclades, the
Dodecanese, and the Ionian island groups.

Today the Cyclades include about two hundred small
islands, but in ancient times the group was considered to be
a rough circle formed by eleven islands surrounding the
sacred Island of Delos. The distinctive Cycladic architecture of
square whitewashed houses set off against sun-baked, rocky
hills seems almost dreamlike. The narrow streets lead
between the houses down to a beach of hard yellow pebbles.
The deep blue water sparkles in the sunlight, while green
islands loom in the distance. Inland the scene changes, as
dark gray stone walls fence off the fields of gray-green
grasses and yellow flowers. On the terraced hills, brown scrub
predominates, while here and there a stone windmill with
spindly masts dominates a peak.

Traveling from island to island, one notes that the ocean
seems alive with a rainbow of colors, often deep blue and

purple, sometimes becoming a lavender mist that makes the
islands seem to float on the water. Overhead fly huge white
pelicans, loved by the island people.

The sun-drenched Greek Islands are an inseparable part of
Western culture and history. Archeological sites abound on
each island, and include everything from bare evidence of
human habitation to the beautiful remains of ancient temples
and theaters, Byzantine churches, and old villages that retain
their original street plans. The Dodecanese Island group
includes the Island of Rhodes, where a gigantic colossus was
one of the Seven Wonders of the ancient world.

Tourists come to enjoy the beaches and to play in the
warm waters, or to party in the local towns. But a lovely and
extraordinary ecosystem is at work in these islands, a natural
world that is missed by most visitors. It is this combination of
rich cultural heritage and exquisite beauty that makes the
islands special.

Corfu is one of the largest of the islands, thirty-three miles
long and fifteen miles wide. Close to the coast of Albania,

Corfu includes rugged Mount Pantokrator, with steep slopes, winding roads, and tiny turquoise coves below. The inland portion of the island includes fertile valleys with citrus and olive groves and a wetland nature preserve at Korisson Lake.

Each of the smaller islands is a delight and has something that sets it apart from the others. For instance, Euboea and Ikaria have hot sulfur springs and medicinal plants. Hydra has no autos, and the town has been preserved as it looked in the 1800s, with donkeys used for transportation. The town of Kaloni on the island of Lesbos is a bird-watcher's paradise and has marvelous stands of wildflowers.

Everywhere there are dynamic cliff faces, some covered with buildings and others standing alone, rising swiftly from the sea and covered with rocks piled up like children's blocks. Some of the islands have ochre-colored earth, while others display red soil between the intense green of the olive trees. Fishing boats with colorful sails mix with pleasure boats on the sea. Most of the islands have their best

beaches on the west side. Some have white sand, while others have pebbles of yellow, black and gray.

There are many small islands, but there are also some large ones. Crete is the largest of the Greek Islands, and the fifth largest island in the Mediterranean, measuring 160 miles east to west and 6 to 35 miles north to south. The beach at Vai has the only palm tree forest in Europe, and western Crete, known as the Kastelli, has lovely small villages and wonderful beaches. Crete has more than its share of archeological treasures.

A dynamic mix of history and nature exists within the incredibly diverse Greek Islands. Despite their popularity with tourists, however, many visitors never see or appreciate the natural delights found here. To the discerning eye, the Greek Islands present incredible opportunities to examine and understand nature. And in the end, each of Europe's natural wonders—from fantastic Iceland to the legendary Black Forest, from the rolling hills of Tuscany to the breathtakingly beautiful Greek Islands—accomplishes this important goal.

88
South of Phuket, Thailand's coastline provides spectacular scenery, with incredible, isolated white sand beaches and lovely limestone bluffs. The Thai coast along the Andaman Sea is one of the most beautiful places on earth.

ASIA

INTRODUCTION

From the depths of the Dead Sea to the crest of Mount Everest, Asia is a continent of superlatives. Not only does it contain the highest and the lowest points on earth, but it is also the largest continent in the world and has the largest population. Asia covers nearly one-third of the earth's land area, has the greatest extremes in temperature of anywhere on earth, the heaviest rains, and the largest arid regions. In terms of physical beauty, Asia ranges from lush tropical rainforests to sandy deserts and from flat, seemingly endless plains to enormous snow-clad mountains. By any measure, Asia is incredibly diverse in its multitude of natural resources.

Throughout the ages, the Asian people have been separated from one another by their continent's geography. Becauses of mountain ranges and deserts that kept them apart, they developed startlingly different cultures and languages. Large numbers of people in Asia live along rivers and on the coasts, while smaller numbers dwell inland in mountain and desert regions. The earliest civilizations and all

the world's great religions, including Buddhism, Christianity, Islam, Judaism, Taoism, Hinduism, Confucianism, and Shintoism began in Asia.

Asia measures about 6,000 miles from east to west, and 5,400 from north to south. It encompasses 17,119,000 square miles and has a coastline of 80,205 miles. Asia extends from the Arctic Ocean in the north to the Indian Ocean in the south, from the Ural Mountains in the west to the Pacific Ocean in the east. It is home to the great civilizations of India, China, Japan, and Tibet.

Asia has six major land regions separated from each other by mountain ranges. Northern Asia, in the region of Russia called Siberia, makes up nearly a full third of the immense continent. It stretches from the Urals to the Pacific, and is divided from the rest of the continent by several mountain ranges, including the Hindu Kush. Endless flat tundra, hardy fir forests, and plains called the steppes, covered with green grasses waving in the wind, form bands running from west to east across the northern part of Asia.

88–89
The topography of Guilin, China, is unlike that found anywhere else in the world.

Nestled along the Li River in southern China, the city is dominated by unusual hills. Beautiful regions such as this

one were created by the erosive forces of water on limestone, which created the dramatically sculpted hills.

89 Bottom left
Midway down Thailand's long peninsular tail near the town of Phuket lies Pang-nga Bay. More than twelve miles wide, the bay is

truly spectacular. Limestone pinnacles like this one rise hundreds of feet from the Andaman Sea. Many of the towers contain caves.

89 Bottom right
The country of Nepal is noted for its spectacular mountains, but few stop to contemplate the lovely valleys cut by many rushing rivers between these stately peaks. Below lie cascading waterfalls, green

temperate forests, meadows carpeted with flowers, and clear blue, icy cold streams that rush and fall from above.

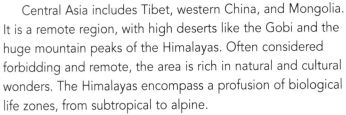

ASIA

INTRODUCTION

Central Asia includes Tibet, western China, and Mongolia. It is a remote region, with high deserts like the Gobi and the huge mountain peaks of the Himalayas. Often considered forbidding and remote, the area is rich in natural and cultural wonders. The Himalayas encompass a profusion of biological life zones, from subtropical to alpine.

Eastern Asia includes Japan, Korea, Taiwan, and the eastern part of China. The area has many hills and mountains and the fertile valleys of the Hwang and Yangtse rivers. Beautiful regions such as Guilin in China were created by the erosive forces of water on limestone, resulting in dramatic sculptures. Volcanic activity created the offshore islands and is part of the Ring of Fire around the Pacific. Gorgeous

volcanoes, most still active, dominate the islands and provide distinctive landmarks recognized around the world, such as Japan's Mount Fuji.

Southern Asia includes Afghanistan, Bhutan, Ceylon, India, Nepal, and Pakistan. Upland plateaus and broad river valleys course through lands that range from dry deserts to lush rainforests. The southern portion of the Himalayas, including the highest mountains in the world, falls within this region. The subcontinent of India has fascinated men for thousands of years with its incredible topography and wealth of natural resources.

Southeast Asia encompasses the area between Myanmar and Vietnam and south to the Malay Peninsula in Indonesia, as well as the the offshore islands, the Philippines. Rolling hills and tropical rainforests characterize the region, which is hot and humid year round. Islands like Borneo provide some of the most exotic and beautiful scenery in the world, both in its jungle and in its underwater wonders. A profusion of plant and animal life makes this region an incredibly rich laboratory for biologists.

Southwest Asia, also known as the Middle East, includes Iran, Iraq, Syria, Lebanon, Turkey, Jordan, Israel, and the countries of the Arabian Peninsula. Important rivers include the Tigris and Euphrates, where the world's earliest civilizations began. The Middle East contains great plains like the one in Iraq and broad desert plateaus like the Arabian Peninsula. The beautiful deserts of the Sinai and its neighbor, the Negev, and the lowlands of the Dead Sea are fascinating areas to explore.

Of course, Asia is also noted for it cultural wonders. Its great memorials and religious buildings inspire awe. Among the treasures found here are Hagia Sofia, the Taj Mahal, the great Wall of China, the Forbidden City, the ruins of Petra, Angkor Wat, and innumerable tea houses and Shinto shrines.

Asia's many natural wonders include towering mountains like the Himalayas, great deserts of sand, lowlands like the Dead Sea, jungles like those found in Borneo, and spectacular volcanoes like Mount Fuji. The spectacular natural wonders of Asia befit the world's largest continent. They are big— sometimes incomprehensibly vast— and as varied as the many peoples who live on the continent.

90 Top right
This view of Mount Everest was taken from space during the summer, so little snow obscures the three dimensional quality of the image. Everest, covered in snow, can clearly be seen at the center of this view, with powerful rivers snaking out away from it in all directions.

90 Bottom
Swathed in mysterious clouds, K2 in the Karakoram Range of Pakistan is the second tallest mountain on earth, rising to 28,250 feet. The mighty walls of these incredible mountains seem to rise endlessly toward the sky.

90-91
At 29,028 feet, Mount Everest is the world's tallest mountain. Its north face is a forbidding and challenging presence in the midst of the mighty Himalaya chain. Known as Sagarmatha in Nepal and Quomolongma in Tibet, Everest is beautiful and mysterious, continually swathed in snow and clouds.

91 Bottom
The Himalayas spread out before the viewer in this photo much as they look to a mountain climber who has reached the crest of one of the peaks. The north face of Mount Jannu in Nepal is seen in the foreground, while the chain of mountains seems to go on toward the horizon indefinitely. Mount Everest, the highest peak on the skyline, can be seen in the distance.

92–93
The landscape near Uchisar in central Turkey is a good introduction to the strange wonders of Cappadocia. Erosion formed the

pilasterlike cones, which seem to back outward from the face of the rock. Uchisar is a region known for its natural rock fortifications carved into the volcanic tufa stone.

92 Bottom left
Along the road between Cavusin and Goreme one sees wonders such as these eroded rocks. Goreme is now an open-air

museum that allows visitors to see the rock-cut churches and their interior frescoes. This road was once the Old Silk Road between the Middle East and Asia.

93 Top
This fairy chimney with three heads is located in Zelve. The Zelve region is known for its underground cities.

93 Bottom
Uchicar is a small town in Turkey carved from tufa formations. Doors and windows can be seen in the formations.

92 Bottom right
At the request of the Turkish Government, families have now moved out of these ancient rock homes and into conventional

houses, making this area near Zelve an outdoor museum. Many of these Peribacalar, or fairy chimneys, began as monastic retreats hundreds of years ago.

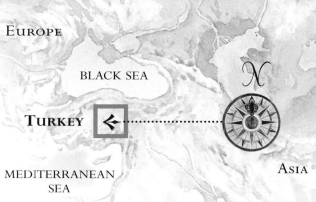

FAIRY CHIMNEYS

CAPPADOCIA, TURKEY

Walking down a dusty path, the rusty pink rock formations rise above you. As you proceed deeper into the valley, the fantastic cones of rock take on various odd shapes, all variations on the basic theme. Some cones include large rocks balanced on their tops, looking as though they might fall at the slightest footstep or breath of wind. Other cones have doors carved into their sides and have been hollowed out for human habitation. Mile after mile of these strange natural monuments, rising abruptly from the earth, characterize the region of Turkey known as Cappadocia.

Cappadocia is located in central Turkey and is roughly triangular in shape, with the points of the triangle sixty miles apart, with Kayseri on the northeast, Aksaray on the west, and Nigde on the south. Cappadocia's lunarlike landscape was formed from a vast volcanic plateau created fifty to one hundred fifty million years ago by the eruptions of the Erciyes and Hasan volcanoes. The volcanoes spouted lava and ash called tufa that formed a thick layer on an ancient Cappadocian lake. Erosion of wind and water sculpted interconnected, scenic valleys and steep canyons of andesite and basalt surrounded by limestone plateaus. The tufa itself, softer than most rock, was shaped by the forces of wind, rain, and snow into incredible shapes resembling cones, pyramids, mushrooms, animals, and columns. These strange geological formations have been called Peribacalar or Fairy Chimneys.

Settlements in this remote region go back more than ten thousand years. The Hittites lived here from 2000 to 1200 b.c., and called it Kizzuwatna, meaning below country. Persian horse breeders inhabited the region from 546 to 333 b.c., and called it Katpatuka, or land of the beautiful horses. The modern Cappadocia is the Greek version of this name. Early Christians also lived here, originally led by St. Paul, and regarded the wild beauty and remoteness of the place as a good spot to fast, pray, and do penance without fear of persecution. They built over one thousand churches cut into the rock at Goreme, Soganly and Ihlara, canyons. Cities were also built underground at Kaymakly, Derinkuyu, Ozknoak, and Mazyroy, with natural rock castles at Ortahisar and Uchisar. These strange homes, forts, and churches could be built because of the nature of the tufa, a soft pumice rock easy to dig into and carve. In some places architectural features like columns, pilasters, and porticoes were fashioned, making the interiors of the dwellings look like they are cut from hard stone.

Walking among the rock formations, one notices how they change color with the shifting light of day. In the springtime the valleys and plateaus are awash with wildflowers and greenery. In autumn the stark contrast of light and shadow on the rocks makes for magnificent views. And winter is unsurpassed for local beauty. Snow often covers and lies among the fairy chimneys, the white contrasting with the light gray and red rock.

The best-known area for viewing the wonders of the region is around the Goreme Valley in north-eastern Cappadocia,

where the terrain consists of a fine-grained, compressed ash. The rock often has a pink hue from the underlying sand bed. Strange sights abound. A three-headed chimney may be seen near a grape vineyard. People may be spotted emerging from a seemingly solid rock pillar. In some places a layer of nonvolcanic rock can be seen balanced picturesquely on an eroded cone.

The western portion of Cappadocia is less explored and contains intriguing formations. The narrow Ihlara Valley is lush compared to other parts of the region. The valley floor in Ihlara is composed of hardened volcanic ash. Throughout the region, volcanic landforms are honeycombed with cut-rock settlements, monasteries, churches, and underground cities dating from the Byzantine period.

Altogether, Cappadocia is a land of weird, fantastic beauty, a harsh land for many years forgotten, one fascinating to explore. It is a land of contemplation and peace, remarkable above all for its mysterious fairy chimneys.

94 Top
The Chapel of the Holy Trinity is located at the top of 7,500-foot-high Gebel Musa, thought by many to be the Mount Sinai of the Bible. Tradition holds that the chapel was built on the spot where God spoke to Moses in the form of a cloud of fire.

94 Bottom
The Monastery of St. Catherine was founded in the sixth century A.D. at the foot of the Gebel Musa. This is thought to be the spot where God spoke to Moses from the burning bush.

SACRED PLACES

THE SINAI DESERT

The high mountain tops are hard, rounded knobs of rock, reaching great heights and then falling away. These knobs seem to continue forever in all directions. Orange-red rock walls contrast with purple blue shadows. As the hot sun beats down, thoughts inevitably turn to the great figures of three of the world's religions, Moses, Jesus, and Mohammed, all of whom walked the nearby desert. They came to the desert to fast, to pray, and to hear the voice of God. The desert's vastness, its unrelenting harshness, and its complete lack of distraction aid contemplation, help resist temptation, and can inspire revelation.

These sacred places compose the desert lands of the Sinai Peninsula. Bordered by the rift valley on Jordan's eastern side, the Sinai is dry and somewhat forbidding, yet around every bend in the dry stream beds or rock formations lies a surprise. The landscape is composed of rough, reddish-brown granite and sandstone with few plants or any

color. Suddenly, however, the seeming nothingness is punctuated by great bursts of color in the rocks or by the lush beauty of brilliant green palm trees clustered around a natural spring.

The desert is a world of incredible harshness, and some find it hard to see its appeal. For others, the desert is one of the most beautiful and peaceful places on earth. The scenery may not be conventional—and to those who love it that is certainly one of its charms—and water may be scarce, but the desert presents a simple, unusual, and surprising beauty.

It is understandably difficult for vegetation to survive in this landscape. The Sinai is not made up of sand dunes. Instead it consists of strangely sculpted rock formations and dry, cracked mountains. Canyons with steep-walled cliffs of brown and yellow rock are marked by hollows and pockmarks so large that a person can easily crawl inside. The land also contains huge boulders the size of houses and oddities caused by erosion. Within occasional oases that are fed by ground water, little areas with greenery of Maidenhair ferns and fig trees live, presenting a sharp contrast to the dry rocks above. Echoes of falling rock, animal movement, and sometimes even human voices seem to come from nowhere and everywhere at the same time. These ghostly sounds help support the incredible legacy of visions, revelations, and deep faith engendered by prophets who once walked here. Since few people live in the Sinai, solitude surrounds you, creating a perfect environment for contemplation. Solitary hikes highlight the area's natural beauty, its unsettling quiet, and the subtle colors that are all about.

To the east is the Great Rift Valley, site of very active faults in the earth's crust. The rift opened as the rock on either side of it split apart, beginning the formation of the Dead Sea. This geological divide began about twenty-five million years ago between the tectonic plates of Asia and Africa. As the mighty continents began to divide, the Red Sea began to form. Deep beneath the Red Sea, evidence exists that Asia and Africa are still drifting apart. Hot 130 degree Fahrenheit water spurts from the underwater fissures between them. To the north of the Red Sea, the rift valley is

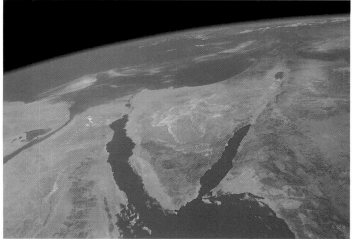

This view from space shows the Sinai Peninsula at the very center. At the bottom of the view is the stunningly blue Red Sea, with its two arms stretched upward.

Inscriptions at Rod Al Air tell stories of ancient man from 500,000 years ago up to fairly recent times. Seen here are hieroglyphs from the ancient Egyptians.

94–95
An oasis at Ain Uhm Ahkmed exemplifies those scattered areas within the desert country where natural springs allow water to bubble up from beneath the earth's surface. Oases are generally tended, and their water used to cultivate crops.

The desert fox is a variation of the common red fox. It feeds on small mammals, especially rodents and birds.

A desert view in the Sinai at Wadi Arada Hakatan includes both flat, rocky terrain and large rock formations.

96 Center
A forest of columns, eroded geological shapes in rock, can be seen in the Sinai Desert near Gebel Fuga. These columns are

but one of the many forms of arresting geological features which make the desert a lively and surprising place.

about fifteen miles wide and is bordered by steep cliffs, with the eastern uplands of Jordan rising to heights of over five thousand feet to the east.

On the west side of the rift valley lies the Sinai Peninsula, a huge 37,820-square-mile triangle of dry landforms. Connected to the Sahara Desert to the west, the sands of the Sinai cover a shield of rock that also makes up the Arabian Peninsula. The southwestern portion of the peninsula is raised, and the southeastern part falls into the Gulf of Aqaba and the Great Rift Valley. This is not a flat and endless desert, but rather a mountain and plateau system as the colorful and beautiful rock formations in the area reveal. Geologically this region is a horst, a horizontal block left standing between two faults, like Germany's Black Forest.

96 Bottom
The Gulf of Aqaba comes to an end on the Egyptian shores near Taba. The stark, barren-looking hills present a sharp contrast to the coral-laced waters of the Gulf, teeming with marine life. The Gulf leads southward and into the Red Sea.

96–97
Rock art inscriptions at Rod al Air indicate the passage of human beings through this region as long as 500,000 years ago. The cradle of humanity may have been Africa, but our adolescence was spent in the desert regions. The lack of water erosion in the desert environment has preserved these pictographs as though they were new.

In the north of the Sinai are flat, dusty plains, while the central portion of the peninsula is composed of a limestone massif topped by the Tih Desert. In the south are the high mountains, the most spectacular and renowned portion of the peninsula. Palm trees line rocky stream beds, which wind in and out of the foothills. The mountains are composed of sandstone of a deep red color, interspersed with gray and yellow. The highest of the Sinai Mountains, Jabal Katrinah, rises 8,750 feet above sea level.

At the eastern edge of the peninsula, Sinai's purple mountains tumble into the Gulf of Aqaba, complete with turquoise waters, red coral, and a huge variety of tropical fish. The gulf supports a dense population of over one hundred species of coral, eight hundred species of fishes, and hundreds of species of mollusks and crustaceans. The coral is incredibly bright and vibrant. Fish are abundant and include crimson-red lunar groupers and bluespine unicorn fish. The wonders beneath the surface of the waters makes this one of the world's most popular diving and snorkeling spots.

Man has lived in these deserts for a very long time. Rock inscriptions on the walls of the canyons date back over 500,000 years. The Sinai Peninsula has been fought over for thousands of years, not because of any inherent wealth but because it was and is the only land route between Asia and Africa. The Sinai Desert is where Moses received the Ten Commandments and saw the burning bush, events still commemorated by area residents, particularly at the Monastery of St. Catherine. The Gebel Musa, three mountains in the Sinai range, have at one time or another all been identified as Mount Sinai or the Mount Horeb of the Bible.

Animal life in the Sinai includes the brilliant white Arabian oryx, the Nubian goat, the desert rat, the gerbil, the black-tailed dormouse, the sand cat, the desert fox, the Arabian wolf, and the leopard. Houbara bustards with white and black feathers perform their mating rituals each spring. Egyptian, Israeli, and Jordanian officials have worked in recent years to

restore the region's abundant indigenous wildlife.

The desert also includes some extremely hardy plants that have adapted to saline soil conditions and limited rainfall. Occasional oases fed by natural springs support green palm or fig trees, and these are striking among the ochre and red desert rocks. In the higher elevations live acacia and even some juniper trees.

Altogether, the desert of the Sinai is full of hidden wonders. Although at first it may seem dry and lifeless, the desert has marvelous areas of tumbling water and unexpected greenery. The impressive rock formations, with their geological variations and dramatic color, have a majesty that is continually rewarding to explore. After experiencing the wild, gorgeous loneliness of this place, it is easy to understand why the most revered religious leaders came here to refresh their spirits and to talk with God.

97
Multicolored rocks on the road to the Monastery of St. Catherine reveal several forms of erosion. Desert rocks tell many stories about the geological history of the area as well as the cultural.

98–99
The southern Sinai is the locale of the Gebel Musa, most often associated with the Mount Sinai of the Bible, upon which Moses received the Ten Commandments from God. The desert is certainly a place of contemplation, where many religious mystics and prophets have gained a greater personal understanding of the metaphysical realm. This photo shows the barren loneliness, wild topography, incredible beauty, and the colors of the Sinai.

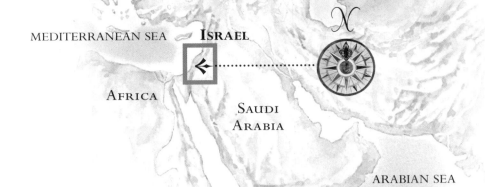

MEDITERRANEAN SEA **ISRAEL**

AFRICA

SAUDI ARABIA

ARABIAN SEA

THE HIGH AND THE LOW

THE NEGEV AND THE DEAD SEA, ISRAEL

Walking through the desert, one begins to search for signs of wildlife, for plants and birds, for a hint of color or a breath of life. And then you begin walking down the dry Tzin riverbed. Scrubby tamarisk trees and other spots of green can be seen. It is near sunset and suddenly beautiful colors seem to light up the sky with tones of fiery red, crimson, and magenta. The rocks around you turn an almost monochromatic landscape into a kaleidoscopic show of color and light. Sunset seems to linger in this country, and one can certainly enjoy every moment of it. This is the Negev region of Israel.

The Negev is a huge triangle of five thousand square

miles of land which composes the southern portion of Israel. It is located north of the Gulf of Aqaba and southwest of the Dead Sea, the most saline body of water in the world. Although spoken of in terms of a desert (Negev means dry land), the Negev is rich in variety of terrain, and portions are irrigated and used as cropland.

Looming high over these shores, the Negev Highlands rise from 1,300 to 3,000 feet above sea level. Due to their altitude the highlands have much milder, cooler weather than the desert below. Today, the region has a small population, which has led to the protection of its various natural resources. Four Israeli National Parks, representing archeological sites of ancient Nabatean cities, are located in the highlands. Near Qumran on the northwest shore of the sea, a young Bedouin boy found the Dead Sea Scrolls in a cave in 1947.

Ein Avdat is a magnificent gorge located deep within the Negev Desert on the riverbed of the Wadi Tzin. Just a little over a mile in length, the narrow and stunningly beautiful canyon is topped with boulders along the cliffs. The Tzin is home to many forms of wildlife, including ibex, sand rats, desert leopards, and rock doves. The riverbed is usually dry, so visitors can walk its length to view thick foliage and incredible rock formations. The area is also famous for deep pools filled by underground streams, formed after thousands of years of erosion and flooding. The quiet, echoing space of the canyons highlights the sound of the wind and the occasional footstep of an animal.

The Negev blooms each spring with wildflowers. Throughout February, March, and April the desert puts on an unrivaled show of color, particularly with irises. Some are chocolate brown, others dark violet and purple. The violet Negev iris grows further south than any other iris in the world on sands in the Negev and the Sinai, while the Yeroham iris lives in one localized area where a nature preserve has been created to protect it.

Much of the desert wildlife in the Negev has been killed off, including oryx, onagers, and ostriches. Other animals, including gazelles, ibexes and, with the reintroduction of their prey, leopards, have been saved through the efforts of conservationists. Biologists at the Hai-Bar Wildlife Preserve have been successful in reintroducing native species in small numbers to the Negev, including onagers and oryx. Desert animals are usually active at night rather than in the hot temperatures of the daylight hours. Other animals that can be seen in the Negev are foxes, hyenas, caracals, and wolves. Millions of birds migrate from Europe and Asia to Africa, and use the Negev as a watering area. These birds include cranes, cormorants, eagles, vultures, pelicans, and storks.

In the center of the region, to the northeast of the Negev, lies the Dead Sea, nine times saltier than the earth's oceans. It

100 Top
Embankments of white earth separate the salty water of the Dead Sea into pools in this view. Like other parts of the planet thought to be devoid of life, the desert near the Dead Sea surprises the visitor with fascinating bird, reptile, and insect life.

100 Bottom
The waters of the Dead Sea are nine times saltier than the earth's oceans. Evaporation causes the salinity, since the extreme heat of the region causes fresh water to disappear, leaving salt deposits behind.

100–101

High above the waters of the Dead Sea, seen in the background, stands the stronghold of Masada on a flat mountaintop. Hebrew for fortress, Masada is located thirty miles southeast of Jerusalem and is a National Shrine. During the revolt of Jewish Zealots against Roman rule from A.D. 66 to 73, a final band escaped to this former Roman palace, holding out for two years against the besieging forces. At the point of capture, all but seven of the 960 Jewish defenders committed mass suicide in A.D. 73.

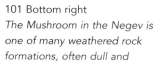

101 Bottom right

The Mushroom in the Negev is one of many weathered rock formations, often dull and monochromatic during the strong light of midday, but alive with color and interest during dawn and sunset.

101 Bottom left

Mizpe Ramon stands 863 feet above the surrounding desert, with the dry bed of the Nahal Ramon running at its feet. Parallel ridges and folds of the earth can be seen in the upper right hand corner, fading into the distance in successive order. Despite the dry nature of the area, the riverbeds are the habitat areas of wildlife and interesting, hardy plants.

102–103
In the Jordanian Desert to the east of the Dead Sea is the incredible Wadi Rum. Wadi Rum features sculpted sandstone monoliths and buttes called jebels, sliced by black canyons. Wadi Rum is notable for its incredible silence.

102 Bottom left
This view from space was taken directly over Saudi Arabia. The picture shows the diminishing elevation, beginning with mountains along the Red Sea at the upper left, running to plateau country near the center, and becoming a flat desert in the Ad Dahna on the right.

is called the Dead Sea because few life forms can thrive in its salty water. The extreme heat of the region causes fresh water to evaporate rapidly, leaving behind salt deposits and a beautiful atmospheric haze of moisture in the air. Due to all this evaporation, the Dead Sea is gradually becoming smaller. Beautiful white salt deposits line its shores, looking like fresh snow stretching off to the purple mountains on the horizon.

The Dead Sea is located below the Judean Desert, which over the course of less than twenty miles quickly descends to the water, ranging from an altitude of nearly three thousand feet above sea level down to 1,310 feet below

sea level. The last step in this incredible plunge is a six-hundred-foot tall cliff that drops directly to the sea. The Dead Sea lies between Israel and Jordan, just fifteen miles east of Jerusalem, with the Jordan River flowing in from the north end. The water is extremely deep, averaging about one thousand feet.

In the deep valley surrounding the Dead Sea can be found gorgeous green banana plantations, and as one travels lower still, water can be seen leaking down the multicolored walls of the rock rising above the valley. Reaching the sea, boulders of many colors can be seen scattered about on the beach. Along the shore narrow canyons are numerous, many with walls rising five hundred feet on each side, with little creeks emptying into the sea. Swallows fly through the air, and the sound of frogs can be heard over the cascading water. Not surprisingly, crusty salt deposits line the shoreline, along with driftwood bleached white by the sun.

The Dead Sea is famous for the health spas and hot springs that line its shores on the Israeli side. Hot sulfur springs are dotted with modern buildings and health resorts. The saline waters are thought to be therapeutic for bathing. A person can easily float on the water of the Dead Sea because of its high salt content.

All in all, a walk along the desert near the Dead Sea almost certainly results in many surprises, from the beauty of the arid rocks above to the calm water below, from the shores of the sea lined with snow-white salt to the area's beautiful and exotic wildlife.

102 Bottom right
Only the wind disturbs the peace of the rippled sand in the Wadi Rum. The region is perfect for solitary hikes and for discovering natural beauty and seeing incredible colors. The region is inhabited mostly by tent-dwelling Bedouins who travel by camel.

103
Wadi Rum is a haven for geologists who wish to examine the Great Rift Valley and the split between the tectonic plates of Asia and Africa. The eastern uplands of Jordan rise from the rift valley to heights of over five thousand feet, and feature rock formations like those in this view.

104–105
This striking view of the Zin Desert shows the region's dramatic cliff and canyon country as viewed from Mizpe Ramon. The striking colors in the rock change with the day's shifting light, becoming especially vibrant at sunrise and sunset. The region's high country is home to ibex and desert leopards, a silent landscape only broken by the occasional echo of a falling rock or the cry of a raptor wheeling high overhead on the thermal updraft from the canyon floors.

106 Bottom left
The morning sun strikes one of the five Amir Lakes, held in place by a natural limestone dam. Sometimes the water spills over the dam into the Band-i-Amir River below. The average altitude of the lakes is 9,800 feet, balanced between the desert below and the snow-bedecked summit of the Hindu Kush above.

106–107
The incredible lakes of northern Afghanistan known as the Band-i-Amir are located high in the mountains 120 miles west of Kabul. The country is rough, and four-wheel-drive vehicles are needed to reach it.

BLACK SEA

CASPIAN SEA

AFGHANISTAN

SAUDI ARABIA

INDIA

ARABIAN SEA

BLUE GEMS OF THE DESERT

THE BAND-I-AMIR LAKES, AFGHANISTAN

Hot and dusty, you ride for hours in a four-wheel-drive jeep, with sand everywhere, so that by the end of your trip you look like a ghost. The country is rugged and monotonous, with jagged red bluffs and little vegetation.

Suddenly the scene is interrupted by an incredible sight—first a glimmer of blue, the glint of the sun glancing off water, and then, right in front of you, a broad lake that emerges from the reddish brown land. Investigating the lake you realize that there are actually five lakes, all an iridescent azure, and all held amazingly in place by twenty-foot-tall natural limestone dams. Each of the lakes looks as though it might spill over these dams or break through them at any moment, but the water is held there in what seems to be suspended animation.

Nestled in the northern portion of Afghanistan between the Hindu Kush and the border with Turkmenistan, the Band-i-Amir Lakes are located high in the mountains, about 120 miles west of Kabul. Of course, this distance is measured as the bird flies, for one must cross the Hindu Kush mountain range to reach the Amir Lakes from the nation's capital.

The average altitude of the lakes is at 9,800 feet, and their deep blue water reflects the barren desert country that surrounds them. There is little vegetation but many find great physical beauty in the color of the rocks and their relation to the waters below. The five lakes drain into the Band-i-Amir River, which, like much of the water in Afghanistan, drains within the country. Only one eastern river, the Kabul, reaches the Indus River. There are five lakes and natural dams in the chain, the Band-i-Kahmar, the Band-i-Gulaman, the Band-i-Aibat, the Band-i-Panir and the Band-i-Pudina. Mineral deposits left over the course of millions of years created the rocks that hold these ice-cold lakes, leaving them as much as twenty feet above the surrounding countryside.

Unfortunately, these wonders of nature are quite difficult to see today, because Afghanistan has been torn by political unrest, which has made the nation a very dangerous place for any outsiders. Like the people and the country of Afghanistan, however, the topography and geology has an ancient heritage and an incredible resiliency. It is almost certain that the Band-i-Amir lakes still exist in the pristine state shown in these views, and that one day they will be easily available for all to see once again.

106 Bottom right
The monotonous, rugged country surrounding the lakes is tempered by the multicolored sparkles of light from their surfaces. There are five lakes and natural dams in the chain, the Band-i-Kahmar, the Band-i-Gulaman, the Band-i-Aibat, the Band-i-Panir, and the Band-i-Pudina.

107 Top
The Band-i-Amir Lakes were created by extremes of climate and geology. Mineral deposits left over the course of millions of years created the rocks that hold these ice-cold lakes, leaving them as much as twenty feet above the surrounding countryside.

107 Bottom
The deep blue water of the Amir Lakes reflects the barren desert country that surrounds them. There is little vegetation in this hot and dusty region, but great physical beauty can be found in the color of the rocks and their relation to the waters below.

KARAKORAM

HIMALAYAS

INDIA

ARABIAN
SEA

THE ROOF
OF THE WORLD

THE HIMALAYAS AND KARAKORAM

INDIAN OCEAN

108
The sharp pyramid of Dorje Lhakpa in Nepal looks almost as if it were constructed by a master builder rather than by the random hand of nature. The Himalayas are the result of two tectonic plates colliding with one another at the rate of about five inches per year. The mountains are pushed upward about one-fourth an inch annually.

The snow crunches beneath your feet, as bright sunshine floods the plateau below the high peaks. Their sharp outlines tower above, seemingly higher than the clouds themselves. The air is thin and crisp, and your breath flows out in torrents of misty spray. Far below, the green of forests and meadows can be seen, while above you the glistening pyramidal heights reflect the light like diamonds. The successive mountains and valleys ahead form V-shaped slashes across your route, accentuating distance and height. Each step brings a more spectacular view as you trudge higher, and higher, and higher. You are headed for the Roof of the World, which is found in the mountains of the Himalayas.

The Himalayas are the tallest mountain range on earth, and their statistics are staggeringly large. The name Himalaya derives from the Sanskrit for Abode of Snow. The rugged, enormous peaks are, as the name suggests, perpetually draped in snow and covered with moving glaciers. Over many centuries the incredible beauty and massive extent of the Himalayas have attracted not only nature lovers but also holy men in search of divine inspiration. Like the wild deserts, these mountains dwarf human concerns while suggesting the power and majesty of God. Religious pilgrims today continue to be drawn to places of quiet contemplation like the mount Kailash and the lake Mansarovar in Tibet, Thyang Boche in Nepal, Bandrinath, Kedamath, Yamunotri, and Gangotri of UttarPradesh, Amarnath in Kashmir, and Hemis in Ladakh. For many westerners, the simple Buddhist monasteries may be reminiscent of *Lost Horizon*, the popular 1935 novel by James Hilton.

Like the Alps, the Himalayas are very young mountains, known as fold mountains because they extend in a series of parallel ridges or folds. The Himalayas really consist of three mighty mountain ranges, the Hindu Kush in Afghanistan, Tajikistan, and Pakistan; the Karakoram in Pakistan and China; and the Himalayas in Pakistan, India, Nepal, Bhutan, and Tibet (China). Like the Alps, all these mountains are the result of two continental plates, in this case the Eurasian and the

108–109
Mounts Makalu and Chimolonzo are reflected in a small Tibetan lake in this view of a glorious dawn in the Kama Valley. Located near Mount Everest, the two peaks are seen looking southward in this photo. Makalu is the world's fifth highest mountain at 27,824 feet.

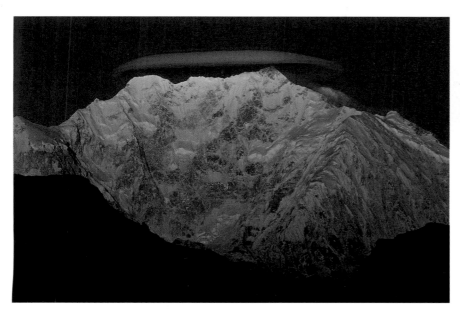

109 Bottom left
Mount Kanchenjunga is 28,208 feet tall and is located in Sikkim, a peninsular section of India that stretches to the Himalayas. The mountain is seen here with the Zemu Glacier, incorporated within Kanchenjunga National Park.

109 Bottom right
The peak of Dorje Lhakpa is seen in this view, showing the serrations of snow. The Himalayas form a barrier to weather systems, which dump their moisture on the south side of the range, leaving the north side extremely arid.

110 Top left
The summit of Broad Peak, the thirteenth tallest mountain in the world, is seen in the foreground of this aerial view, while the north peak of the same mountain is seen behind, with the pyramid of K2 popping above the clouds in the left background.

110 Top right
Jagged spikes of the Karakoram Range near K2 are seen in this magnificent view. The Karakoram Range consists of regular cones of ice-covered limestone on a granite base. Karakoram encompasses a total of sixty peaks that average about 22,000 feet.

110 Bottom
A full shot of K2 epitomizes the power and majesty of this mountain. Also unofficially known as Mount Godwin-Austen after the second European to see the peak, the name has never taken root. The locals call the peak by many names: Chogori, Lambha Paha, Dapsang, and Kecho—K2.

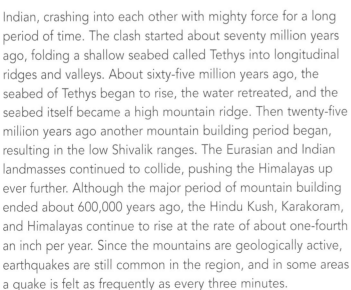

Indian, crashing into each other with mighty force for a long period of time. The clash started about seventy million years ago, folding a shallow seabed called Tethys into longitudinal ridges and valleys. About sixty-five million years ago, the seabed of Tethys began to rise, the water retreated, and the seabed itself became a high mountain ridge. Then twenty-five million years ago another mountain building period began, resulting in the low Shivalik ranges. The Eurasian and Indian landmasses continued to collide, pushing the Himalayas up ever further. Although the major period of mountain building ended about 600,000 years ago, the Hindu Kush, Karakoram, and Himalayas continue to rise at the rate of about one-fourth an inch per year. Since the mountains are geologically active, earthquakes are still common in the region, and in some areas a quake is felt as frequently as every three minutes.

The combined mountain ranges stretch for 1,550 miles from west to east, with an average width of 125 to 250 miles. They cover 229,500 square miles at an average height of 20,000 feet. The three parallel ranges contain more than thirty peaks higher than 24,000 feet, and form the northern border of the Indian subcontinent. Even the passes through these mountains are higher above sea level than Mont Blanc, the tallest of the Alps. The world's highest peak, Mount Everest, stands in the range at 29,028 feet, followed by K2 at 28,250 feet and Kanchenjunga at 28,208 feet. The Great Himalaya Range forms the backbone of the system and includes nine of the fourteen highest peaks in the world. Nineteen major rivers drain from the Himalayas, including the Indus, the Ganga river, and the Brahmaputra. The Hindu Kush, Karakoram and Himalayas fall within the borders of Burma (Myanmar), India, Nepal, Bhutan, Tibet (China), Pakistan, Tajikistan, and Afghanistan.

110–111
On the border of Pakistan and China stands the mighty mountain known as K2, second

largest in the world. Here, the deep ice of a glacier and the light appearance of snow on its surface temper the harsh north

face of the pyramidal monolith. Two Italians, Achille Compagnoni and Lino Lacedelli first climbed K2 in 1954.

But for those who think that the Himalayas are only somber, forbidding mountain peaks perpetually frozen in snow, the region comes as a great surprise. Below the peaks lie cascading waterfalls, green temperate forests, meadows carpeted with flowers, and clear blue, icy cold streams that rush and fall from above. A walk in the lower reaches of the Himalayas reveals the unique character of the region, apparent not only in its geographic features, but also in its people and architecture. Small villages predominate. Shepherds tend flocks of such animals as sheep, goats, and yaks. Well-developed trails, hundreds of years old, connect these villages for trade and for religious activities. Today the trails have become quite popular with tour groups, or trekkers, who plot routes by which they can see some of the region's spectacular scenery without scaling the enormous mountain peaks.

The Himalayas and Karakoram represent an incredible biodiversity due to the enormous height of the mountains and the number of climatic zones. There are two main types of tropical forests in the Himalayas, rainforests and deciduous forests. The rainforests thrive below 3,900 feet in altitude, above which deciduous forests of teak, oak, magnolia, and sal trees take over. These tropical deciduous forests extend up to an altitude of 7,220 feet. Above these timberlands stand the temperate forests of cedar, birch, hazel, maple, and spruce, rising to 8,860 feet. An alpine forest of juniper, rhododendron, mosses, and lichens succeeds the temperate level and extends up to a height of 11,800 feet, where open alpine meadows succeed it. Above 16,400 feet the forests can no longer grow, and they finally give way to alpine grasslands, meadows, and scrublands. At 18,000 feet the permanent snow line begins. Vegetation varies not only

112–113
Mount Everest, the world's tallest mountain at 29,028 feet, greets the first light of day in this view. Since it was first climbed in 1953, hundreds of climbers have duplicated the feat of Hillary and

Norgay. The youngest person was nineteen, the oldest sixty. Two Sherpas have climbed the mountain five or more times, and one man, Yuichiro Miura, actually skied down the slope in 1970.

112 Bottom
The north face of Mount Everest is seen in this view taken in Tibet. The mountain is challenging, and over ninety-five people have been known to lose their lives trying to ascend it.

Today the climb can only be made with government permission, yet more and more people are successful in reaching the top. Money and fitness seem to be the keys to reaching the top of the world.

according to altitude, but also depending upon the side of the mountains one is on. Tropical rainforests thrive in the eastern Himalayas, while dense subtropical and alpine forests are more predominant in the central and western Himalayas. The Transhimalaya region supports only desert due to the rainshadow of the enormous peaks. And the mountains divide the lush, rain-soaked foothills of Nepal in the south from the arid Tibetan plateau in the north.

The beauties of the Himalayas can be enjoyed n many places, perhaps epitomized by the Annapurna region of Nepal with its breathtaking scenery and dramatic ceep valleys. The Annapurna Conservation Area is run by a non-profit trust to promote conservation and more responsible tourism. The advent of eco-tourism and renewed interest in the Himalaya region has meant the advent of increased use of the area. This influx of visitors has put a great deal of stress

time. Attempts to climb Everest began in 1922, but were not successful until May 29, 1953, when New Zealander Sir Edmund Hillary and Sherpa guide Tenzing Norgay reached the summit. Since then, over 550 climbers from twenty countries have successfully reached the top of the world, while many have died in the attempt. The climb is extremely dangerous, particularly because of the thin atmosphere and lack of oxygen at this elevation, and can only be undertaken

114 Bottom left
Paiju Peak in the Karakoram Range reveals the strong geological upthrust that created its craggy sides.

114 Bottom right
Mysterious and clouded Trango-Tuomo juts sharply upward from the Pakistani plains. Jagged, ice-crowned peaks like these cap the entire horizon of northern Pakistan.

115
The west wall of Gasherbrum IV in Pakistan's Karakoram Range displays the more irregular nature of these peaks. The mountains seem to be thrust straight up out of the ground and then continue to reach for the sky. The entire range still rises at the rate of about one-fourth an inch each year.

with government permission from Nepal.

The second highest peak in the world is K2, located in Pakistan and about 250 yards smaller than Everest. The world's two highest mountains look very much alike in their massive and dominating appearance. Below Nepal's Kanchejunga, the world's third highest mountain, the windswept, rugged terrain provides awesome views of the towering peaks of the massif under a clear blue sky. Kanchejunga turns to gold and purple in the waning light of day and seems to float above the lower blue clouds of the valley.

There are many other beautiful peaks in the Himalayas, many not widely known because they are literally overshadowed by Everest, K2 and Kanchejunga. But all of these peaks would be giants if they were moved and placed anywhere else on earth. Many of them are far more physically beautiful than their taller neighbors. Himalchuli, for instance, is really a ridge with three main peaks and it has a striking appearance when seen from the west. The deep white snow accentuates the gorgeous blue tones of the serrated mountainside. Annapurna is a wonderful fantasy of a mountain with a strong, rounded central peak and beautiful moods that change with the changing light of day. Pumori is a classic white pyramid in shape, sharply pointed and perfect in appearance. Bhagirathi Parbat, with its three major peaks, dominates the end of the valley leading up to the Gangotri

116 Bottom right
The image shows another of the peaks close to Bhagirathi River in Uttar Pradesh. There are three peaks that stand at the end of the valley leading to Gaumukh.

116–117
The majestic face of Mt. Kailash dominates this view. The mother of many Indian rivers, the Himalayas ironically began as sediment deposited by prehistoric rivers millions of years ago. They were raised by the collision of continental plates and have grown so that they are now the world's tallest range.

116 Bottom left
Gaumukh in northern India is at the end of the Gangotri Glacier, the source of the River Ganga. Nineteen major rivers drain from the Himalayas, including the Indus, the Ganga River, and the Brahmaputra.

The Karakoram range contains a total of sixty peaks averaging 22,000 feet, topped off by the incredible K2 at 28,250. Jagged, ice-capped peaks crown the horizon, but the visitor should not be fooled into thinking that the region is perpetually frigid. Actually, the greater Himalayas are at the same latitude as central Florida and Cairo. Consequently summers can be quite hot, and temperatures descend only as one ascends in altitude.

The people of the Himalayan region take the beauties of the mountains in stride but never take them for granted. It is westerners who seem obsessed with naming each peak, each pass, each stream, and charting it all on maps. How strange

Glacier, the largest glacier in the Himalayas and the source of the Ganga River. Machhapuchhre (fishtail in Nepali) is one of the world's most famous and elegant peaks. It rises to a sharp point, looking almost as though it has a white chimney on the top. A similar peak is Shivling, which has been called the Indian Matterhorn since it resembles that European peak.

Serene, white-faced Nanga Parbat in Pakistan (tenth tallest mountain peak in the world) is the opposite of the sharp-spired mountains of the Himalayas, and looks like a large, massive sail spread out along the horizon. In Pakistan one travels into the Karakoram Range, part of the same Asian spine as the Himalayas but one with its own characteristics, beauty, and grandeur. The serrated ridges of the mountains can be seen along the incredible Karakoram Highway, which winds for eight hundred miles through high, cold mountain passes and around some of the tallest peaks on earth. Between towns one can observe the gray rock that has fallen down the slopes, piling up at the bottom like ashes that have been cleaned out of a fireplace. Above, the bright sunlight glints off the sides of the brilliant white mountains, blinding the viewer at times with its reflected brilliance. When one sees such magnificent peaks, it is easy to understand why the ancient peoples of the region worshiped the mountains and why they are still held in high regard today.

For the locals, the mountains are the abode of gods and saints, a place of pilgrimage. In their snows arise the rivers that bring life and in their massive size is the key to the area's weather.

117 Left
A torrent of the Ganga River, formed from the snowmelt of the Gangotri Glacier, is seen in this photo. The life of India, its rivers, comes from the Himalayas.

117 Right
One of the peaks of the Bhagirathi Group, Garwhal-Shivling is located at the headwaters of the Ganga River in India, west of Nepal.

118 Top
The graceful lines of Mount Siniolchu in Sikkim, India, are tinted red by the light of the sunrise and tempered by the pure white snow in which it is mantled. It can truly be said that each of the mountains of the Himalayas has its own personality Each has an individual look.

Englishman T .G. Montgomerie must have seemed in 1856, as he moved into the region with his surveying equipment, paper, and pencils. Montgomerie is the man who named K2, simply numbering it as one of thirty-five summits in the Karakoram Range that he was charting. The name, or lack thereof, stuck, however, and the world's second tallest mountain is still known by a letter and a number only. To the locals, the mountains were and are all one, all part of the same earth and sky, all sacred.

At the far western edge of the Himalayan ranges lies the Hindu Kush, with twenty-four peaks averaging 23,000 feet. The Hindu Kush is surmounted by Mount Tirichmir, 25,230 feet tall. The interlocking chain of mountains, valleys, rivers, and glaciers makes up an entire region of the world, a region dominated by physical beauty and known for the resilience and physical toughness of its people.

It is not surprising that Tenzing Norgay, a Nepalese guide, shared the honor with Sir Edmund Hillary of being the first to reach the top of Mount Everest. It is equally understandable that Norgay did not want the spotlight. Norgay lived with the mountains every day of his life, a dedicated climber who loved the peaks of his homeland yet took them in stride in a way that an outsider would find difficult.

When considered altogether, the incredible natural resources of the Himalayas present a compendium of most of the world's geographical features—mountains, rainforests, deserts, and plateaus. They are incredible works of nature that remind us of our own relative insignificance in the course of geological time. The natural beauty of these peaks certainly provides anyone who visits them a fitting place to contemplate the wonders of our world.

118–119

The face of Annapurna rises 26,504 feet. The Annapurna area is protected by a Nepalese National Park and is one of the most popular trekking regions in the Himalayas. The famed Annapurna Circuit takes hikers through many types of geological and biological zones with great cultural and natural diversity. A circuit usually takes from two to three weeks.

118 Bottom
Another of Nepal's gorgeous peaks, Chulatse's snowy face and glaciers are seen in this view. The rocky spine of Asia is composed of hundreds of sublime peaks like this.

119 Bottom
The serrated crest of Annapurna heralds breathtaking scenery and dramatic valleys. The Tatopani Gorge that borders the mountain is the deepest in the world at over six thousand feet. The Annapurna Conservation Area that surrounds the peak tries to promote enjoyable tourist experiences while at the same time preserving the natural beauties of the area.

122 Top
Looking like the branches of a tree, the tributaries of a river lead to Sadakan Bay in Borneo. The thick vegetation here is a mangrove forest, an ideal breeding habitat for many types of birds and animals. Mangroves grow in shallow and muddy salt water and have arced, tangled root systems.

122 Center
A female orangutan with her baby faces an uncertain future due to logging operations in Borneo. Male orangs stand four-and-a half feet tall and have an enormous arm span of seven feet. Females are roughly half this size. Orangs live in trees and eat leaves and fruits, particularly figs.

122 Bottom
A Sumatran rhinoceros browses in the thick foliage of the rainforest. Another animal whose range is shrinking quickly due to deforestation and hunting, it is one of five rhinoceros species left in the world. It is differentiated from the others by the combination of two horns, incisors and canine teeth, and an armor-plate skin arrangement. Rhinoceros are herbivorous creatures with poor eyesight but a highly developed sense of smell.

123
An aerial view of the rainforest at Sabah, Malaysia, gives an idea of the seemingly endless profusion of trees and waterways. Low-lying, forested swamps like these protect elephants and orangutans.

BORNEO

the richest coral reefs in the world, harboring barracuda, buffalo fish, jack fish, butterfly fish, parrot fish, clown fish, dragon fish, white tip sharks, giant clams, feather starfish, sea urchins, and many species of turtles. The brilliant coral itself forms a paradise of shapes named staghorn, mushroom, cabbage, and brain, in staggeringly bright yellows, greens, reds, blues, and purples.

Back on dry land, Borneo's wonders continue in the Gunung Mulu National Park, a region created about five million years ago when the uplifting of an ancient ocean floor formed its limestone mountains. Erosion has produced an array of pinnacles amid the green forest, while below is an extensive system of limestone caverns. Wildlife includes colorful hornbills, giant flying squirrels, and flying frogs, along with hundreds of species of butterflies.

Bako National Park on Muara Tebas, a ten-square-mile peninsula jutting into the South China Sea, is a region where waves have eroded rocky headlands, sea arches, and steep sandstone cliffs. Giant monoliths covered with vegetation tower above long stretches of white sand beaches. The jungle supports long-tailed macaques, silver-leaf monkeys, monitor lizards that can grow up to eight feet long, plantain squirrels, wild boars, and mouse deer. The proboscis monkey may be the region's most interesting and reclusive resident. This large primate has a long pink nose, fat belly, thick white tail, gray legs, and an orange back, and is endemic to Borneo. Carnivorous and rare pitcherplants, sundews, and bladderworts can also be found in these forests.

Borneo still has, at least for the present, a wild, spectacular jungle full of fascinating creatures and plant life. Although there are other rainforests in the world, there are few as undisturbed and continuous as those in Borneo. The low-lying forested swamps that protect elephants and orangutans; the jungles with edible ferns, bamboo shoots, and unusual tropical fruits; and the skies alive with egrets, Oriental darters, swifts, and vividly colored bee-eaters make Borneo like nowhere else on earth. This wildlife, along with its coral reefs, caves, and jungles make Borneo one of nature's most spectacular wonders.

126 Bottom left
The Gorges de Diosso near Pointe Noire on the Zaire coast includes this incredible band of uplifted rock forming a ridge in the midst of the forest.

126 Right
A water hole in Tanzania is a place where animals can acquire the water they need for survival, yet it is also a place of danger. Prides of lions sit lazily in the shade of copses of trees, waiting to spot a weak zebra or gazelle.

126 Top left
Unusual trees and plants live within Tanzania's Ngorongoro crater. Ngorongoro is over eleven miles across, with walls rising two thousand feet straight up from the crater floor.

126 Center left
The Serengeti-Mara-Ngorongoro ecosystem of Tanzania covers an area of over 18,000 square miles and provides habitat for wildebeests, lions, cheetahs, giraffes, elephants, and monkeys, totaling roughly over three million creatures. The Serengeti consists of vast areas of flat, open plains and rocky foothills. There are dry regions and patches of lush vegetation around rivers and streams.

127
The cone of active Ol Doinyo Lengai volcano stands 9,442 feet above the Tanzanian plain. Meaning Mountain of God in the Masai language, Ol Doinyo Lengai had its last major eruption in 1966. Lengai erupts lava that has the composition of baking soda. The steep slopes of the mountain are scored with deep cracks where heavy rains have gouged the soft ash.

128–129
The beautiful island of Zanzibar lies off the west coast of Tanzania in the Indian Ocean, a green land of clove and cinnamon trees surrounded by gorgeous beaches and by coral reefs.

Most of the high mountains of the continent are in the northwest and center-east, and include the Atlas Mountains, Mount Kilimanjaro, and Mount Kenya. Most of Africa's mountains are not part of a range, but are a short series of peaks that rise above the plains and rainforests. The Ruwenzori and Virunga Mountains in equatorial Africa are noted for their vegetation and greenery. It is in the Virunga Range that the gentle and reclusive mountain gorillas make their home.

One of the most fascinating aspects of the African continent is its vibrant wildlife. Herds of grazing animals like zebras, antelope, and giraffes migrate on a yearly cycle, while predators like lions, leopards, hyenas, and jackals survive by culling the sick, old, and weak. Majestic elephants and rhinoceroses roam the bush, while hippopotamuses wallow in shallow waters. Leading Africans have been intelligent stewards of their natural resources, establishing national parks and preserves throughout the continent's more than fifty nations. Many of these areas, like the Serengeti and Ngorongoro Crater,

have also received World Heritage status for it is there that Africa's ancient ways are preserved, with majestic animals running free in their natural environment. Neither the plains of North America nor those of Asia can boast such an untouched expanse of their original habitat. Other ecosystems have also been preserved, although they are under more stress from growing populations. Primates like chimpanzees, gorillas, and monkeys live in the rainforests, many of which are protected in parklands. The deserts have their own forms of specialized wildlife that have adapted to a world with little water and comprise some of the most intriguing creatures in the world. Since the desert is so inhospitable to non-natives, these animals have been little affected by the changes of the twentieth century.

The continent is as varied in its plant life as in its animals. Tropical forests are the home of oil palms, fruit trees, and hardwoods like ebony and mahogany. Acacia, baobab, thorn bush, and palm thrive on the wide grasslands. Even the deserts have plants that have adapted to their harsh conditions.

Through much of Africa runs the Nile River, providing life and continuity to the continent. The longest river in the world, the Nile brings water to the largest desert on earth. Its blue waters course past dry desert sand and rock, yellow and white in the glaring sun. Its brown floodwaters bring alluvial soil to parts of the Nile Valley upriver from the Aswan Dam. In historic times, it was these floodwaters that spelled success or failure, life or death for the Egyptian harvest and society.

Africa is a magical place, a continent with resources that can transport us back to the earliest eras and give us insight as to who we are as a species and where we are headed as we begin the twenty-first century.

THE SAHARA DESERT

138 Top left
Desert grasses can be seen in this view of the Marzuq dunes in southern Libya. Light and shadow paint gorgeous pictures on these hills of sand, unfortunately seen and appreciated by very few humans.

The Sahara's desert climate was established over five million years ago during the Pliocene Epoch. Since that time the Sahara has been subject to alternating dry and humid conditions, which have contributed to the unique climate of today's desert. Just eight thousand years ago, the Sahara was a fertile region, with rivers abounding in fish and plains with grazing elephants and giraffes. About 4000 B.C. the climate began to change and the region began to dry up. For the past two thousand years, the climate has been consistently dry, except between the sixteenth and eighteenth centuries, what is called the Little Ice Age in Europe and the Americas, when the amount of precipitation increased slightly.

Today, the Sahara is one of the driest regions in the world, with nothing but gullied hills sculpted by the winds. The Sahara's climate is dry and subtropical in the north, with high summer temperatures of up to 135 degrees Fahrenheit, cold winters, and two rainy seasons. A second, dry tropical climate in the south consists of dry winters, a hot dry season, and a rainy season. Rainfall averages five to ten inches a year, but in some years there is no precipitation at all.

138 Top right
The dunes of the Marzuq region of southern Libya epitomize much of this hot and arid land. The desert climate is believed to have begun about five million years ago, during the Pliocene Epoch. The Sahara's climate has remained very consistent during the past two thousand years.

138 Bottom
Animals live in the southwestern deserts of Libya as well. They include desert foxes, gerbils, jerboas, Cape hares, and dorcas gazelles. The Sahara also supports more than three hundred species of migratory birds.

Despite the dryness of the air and the land, there is water in the Sahara. Like many of the desert's beauties, the water is just hidden subtly from view. Underground aquifers, thought to date from the Pleistocene Epoch, lie beneath much of the Sahara. Many of these aquifers are saucer-shaped artesian basins sandwiched between layers of impermeable rock. Rain soaks down through their rims, and may be stored there for hundreds of years. Some of this water may be under pressure, as in an artesian well. Underground rivers also exist, and surface at the many oases of the desert.

Surface riverbeds exist in the Sahara, although they are more often than not dry. In the north, the greater part of the water flows from the Atlas Mountains and highlands of Libya, Tunisia, Algeria, and Morocco in the form of streams and wadis, called oueds in the Sahara. These dry streambeds support many types of shrubs and grasses. Tall date palms grow along the riverbeds, waving their green fronds in the breeze against the deep blue sky. The mud houses of Berber people can be found in and among the stands of trees in western Africa, their doors painted in bright geometric shapes to keep evil spirits from entering. As one walks away from the oueds and the oases, the landscape becomes otherworldly, seemingly devoid of life, without a single visible living thing, animal or plant.

The Sahara is sparsely covered with various types of vegetation ranging from grasses, shrubs, and trees in the highlands, such as cypress, olive, acacia, and artemisia, to grasses found on the plains including eragrostis, panicum, and aristida. Saline tolerant plants called halphytes can even grow in the geological depressions. All of the Sahara's vegetation has adapted to unreliable precipitation and excessive heat. At oases, date palms grow near the water source—the larger the source, the greater the number of trees.

Walking through the desert one is impressed with the silence. The only sound is one's own footfalls clunking against the rocks and stones. The desert seems to stretch on without limit, and the horizon that surrounds you is the same for 360 degrees. Dead. Barren. Lifeless. Endless. A walk through the Sahara is a walk between oases, hoping one has enough water to make it, whether one is on foot, on a camel, or in a motor vehicle. In the distance one sometimes sees palm trees that break the monotony or perhaps just the pastel rocks of distant mountains or rock formations. The stones on the ground at your feet are gray with rough edges, surrounded by dry, coarse sand.

Larger rocks are dark, some even black, evidence of the volcanism that created much of the region eons ago. You see no animal life, for most animals do not come out into the heat of day.

But the animals are there. They include Fennec foxes, gerbils, mice, jerboas, Cape hares, long-eared hedgehogs, dorcas gazelles, olive baboons, spotted hyenas, common jackals, Ruppell's foxes, striped weasels, and slender mongeese. The Sahara also supports more than three hundred species of migratory birds. Ostriches, secretary birds, guinea fowl, Nubian bustards, desert eagles, and barn owls are year-round residents of portions of the Sahara. The Saharan lakes and water pools support reptiles like frogs, toads, and crocodiles, while the rocks and dunes, harbor lizards, chameleons, skinks, and cobras. Most animals make their homes in the mountains, where crags provide relief from the sun and protection from predators. Looking over the barren landscape, it is difficult to believe that so much life exists here, hidden from view.

After a long trek across the desert sands, following compass and map, the long-awaited oasis appears. You have to hope

138–139
A gorgeous, light brown expanse of desert sand spreads out from a small oasis in the Fezzan Region of southwest Libya. One of the largest nations in Africa, 90 percent of Libya is covered with desert, either in the form of a rocky plain or with sand dunes such as these.

139
The wind ripples and carves gentle ridgelines in the fine white sand of southern Tunisia. The date palm, found only where there is water at an oasis, is the most obvious type of plant found in the desert. It is unlike its more spectacular cousins like halphytes, which have adapted to saline conditions, miniscule precipitation, and intense heat.

140–141
A view from space of northwestern Algeria, not far from the border of Morocco, illustrates the diversity of topographical features in the Sahara. The Kahel Tabelbala combines low hills, dry riverbeds, and the Erg Er Raoui, waves of sand dunes sculpted by the winds. Oxidized (rusted) rock covers the desert in a thin coat, creating a profusion of colors.

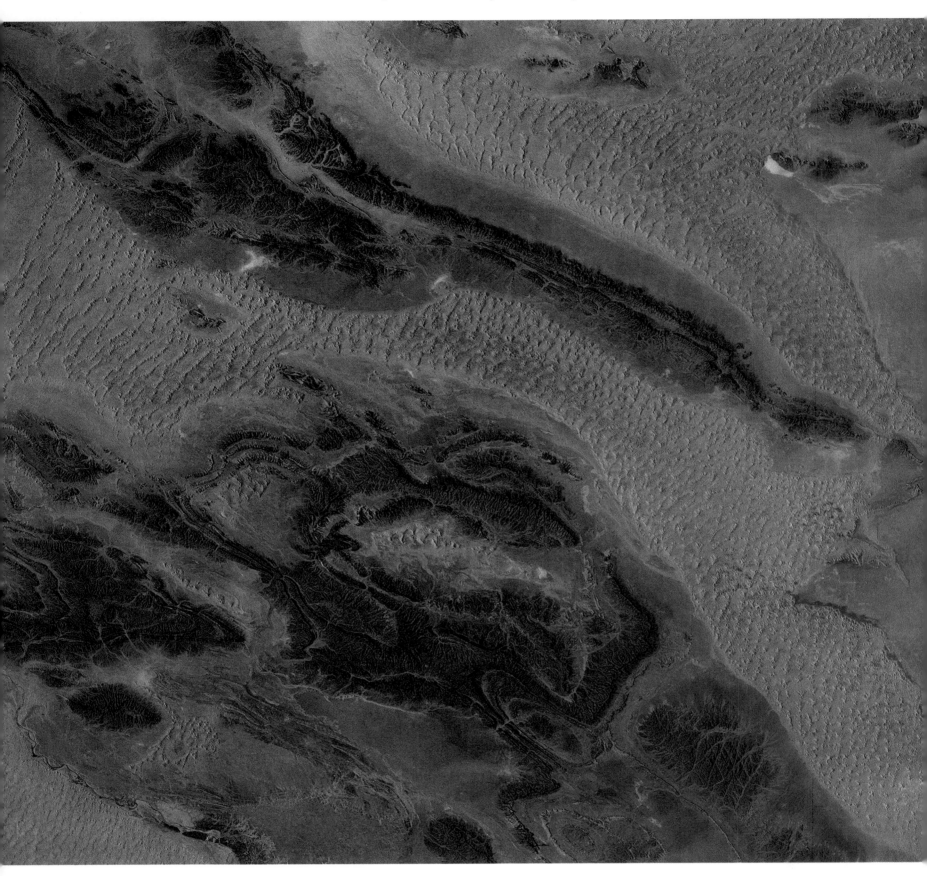

that the well is not dry. Oases with water are havens of calm and tranquility, with shade in and among the palm trees. Oases are like islands in the middle of the huge barren ocean of the desert, where people live, trade, and grow crops. Never do trees look as inviting as they do at an oasis: the lovely brown of the bark and the lush green of the leaves are welcome signs of life in the middle of the desert expanses.

Leaving to continue your trek, you notice the slanting rays of the sun on the crisply defined ridge of a sand dune, snaking in an S-curve toward the horizon. The regions of dunes, or ergs, are crisply defined in the Sahara. In crossing Algeria one would trek over the Great Western and Great Eastern ergs, covering an area about the size of France. It is this scene, the romantic view of camel caravans crossing endless stretches of gigantic dunes, that forms the popular conception of the Sahara.

In the rocky desert mountains of Libya and at Tassili-n-Ajjer, one can find evidence of ancient occupants of the desert in cave and wall paintings. The cool red wall canyons wind into the recesses of the massif, inviting visitors. The Tadrart and Akakus ranges provide stunning landscapes of craggy rock faces red and white in the light of day. The smooth orange sand, blown into little wavelets, pushes right up next to the rough rock towers and pillars which stand about it. This is the essence of the Sahara—miles of sameness punctuated suddenly by extraordinarily beautiful differences in terrain and vegetation.

At the southern edge of the Sahara is the Sahel, a strip of land that separates the desert from a savanna. The word Sahel is derived from the Arabic word for edge or border. The Sahel is

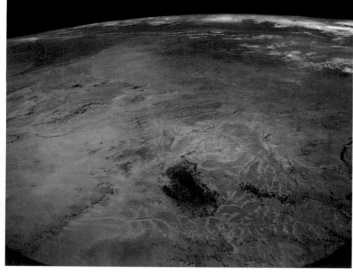

141 Top left
This view from space details the topography of northwest Sudan, a fascinating potpourri of various types of terrain. Earth with a reddish tone is richer, more ancient soil. The dark brown dots at the center are a mass of what may be volcanic cinder cones. Yellow tones are dunes.

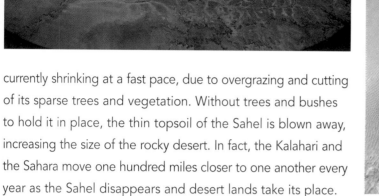

currently shrinking at a fast pace, due to overgrazing and cutting of its sparse trees and vegetation. Without trees and bushes to hold it in place, the thin topsoil of the Sahel is blown away, increasing the size of the rocky desert. In fact, the Kalahari and the Sahara move one hundred miles closer to one another every year as the Sahel disappears and desert lands take its place.

Although seemingly inhospitable to life, the Sahara Desert hides water, plants, and animals. Its sheer size and the incredible diversity of its resources make it one of the world's wonders, although its subtle beauties are an acquired taste. But one can never forget the romance of a desert night, the stars twinkling above and the moon rising to light a landscape like no other on earth. It is there, in the details, the differences and the hidden wonders that the true beauty of the Sahara can be found.

141 Top right
Looking straight down at the Sahara from space, the viewer can see the southwestern corner of Libya with the famous Marzuq Sand Sea. The rugged land at the bottom right is the eastern tip of the Tassili-n-Ajjer Mountains and Tadrart Plateau in Algeria. Small areas of sand dunes are interspersed with the rock outcrops running through the center of the photo.

141 Bottom
In the center of the West African country of Mauritania is an unbelievable twenty-five-mile wide, three-hundred-foot deep natural feature, seen here from space, known as the Mauritania Bull's Eye. How it was created is not understood. Winds eroded the different layers of rock at various rates, creating the effect of rings and ridges.

142–143
Located on the southern tip of the Sinai Peninsula, Ras Mohammed National Park's towering cliffs continue beneath the surface, plunging to Shark Reef's crystal clear waters. Along the underwater portions of these cliffs are soft corals and large schools of fish.

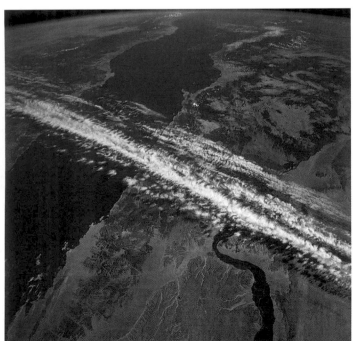

142 Bottom left
The expanse of the Red Sea can be seen in this view from space looking southward. The tip of the Sinai Peninsula can be seen at the lower left edge of the photo, with the Nile River paralleling the Red Sea on the right. The view contrasts the dry desert and the beautiful blue waters of the sea and river.

142 Bottom right
The Tiran Reefs, located near Sharm El Sheikh off the east coast of Sinai, are seen in this aerial view looking north. At the top of the photo is Jackson Reef, known for its diving and its sharks, turtles, and other big fish. Moving south from Jackson, one can also see Woodhouse Reef, Thomas Reef, and the largest, Gordon Reef.

THE WORLD BENEATH

THE RED SEA

You sit on the rail of the boat, hand on your face mask, mouthpiece in place. Your wetsuit fits snugly and your heart beats in anticipation. You are about to jump into the warm waters of the Red Sea at the southern end of the Sinai Peninsula, one of the most renowned diving locations in the world. You let yourself fall backwards into the water, and a cloud of bubbles rises toward the rapidly receding surface. Regaining control of your movements, you flip over to look downward, and at first you think you might be dazed by the concussion of hitting the water. The scene unfolding before your eyes is literally unbelievable. The towering cliffs you saw above the waves, so brown and barren, continue beneath the surface to fade into the deep blue. Coral of every color, seeming to be lit from within, spreads out before you in an incredible tour de force. Large schools of fish gather in the often strong currents with larger fish approaching to feed on them. Finding yourself in the midst of this incredible scene, you reflect on the glorious beauty of the Red Sea.

The Red Sea is famous as the body of water said in the Bible to have been parted at the command of Moses. It was through the Red Sea, most probably the Gulf of Suez, that the Israelites walked on the dry sea bed. When the Egyptian Pharaoh's troops pursued the fleeing Israelites into the abyss, Moses commanded the waters to return to normal, drowning the enemy host.

The surface area around the Red Sea is desert, but this expanse of rocks and occasional vegetation has witnessed the rise and fall of many civilizations over the course of thousands of years. The dark hilly terrain on the west shore of the sea is the Arabian Desert or Eastern Desert of Egypt, and the low mountains along the eastern shore are the Al Hijaz Mountains in Saudi Arabia.

The Red Sea, and particularly the world beneath its waves, is one of nature's true wonders. The waters are saltier than most seas because the region's high winds and extreme heat cause the moisture to evaporate into the air rapidly. The Red Sea is known for its strong currents, currents that can be particularly dangerous to divers. The sea's famous coral reefs are concentrated along the western coastal waters. The plant and animal life of the Red Sea is different from that of the Mediterranean, and includes sharks, stingrays, turtles, dolphins, colorful coral, sponge, starfish, and mollusks.

The Red Sea is an arm of the Indian Ocean that separates the Arabian peninsula from northeastern Africa. The northern

143 Top
The sandy beaches of the Daleh Islands indicate the physical beauty of the Red Sea region above the water. The Red Sea extends 1,400 miles and includes beautiful coasts in Egypt, Sudan, Ethiopia, Yemen, Eritrea, and Saudi Arabia, which in some areas are desertlike.

143 Bottom
An incredible burst of color explodes as bright orange goldfish (anthias) swarm near soft red and pink coral. The clear blue waters make it hard for divers to determine depth, and many nearly exceed the limits of their tanks before returning to the surface.

Coral form in wonderful, branchlike configurations beneath the Red Sea. With a skeleton, these are considered to be hard corals. Far more delicate than they appear, coral can be easily destroyed by lack of light, pollution. Even human perspiration in the water can pose a threat to these beautiful organisms.

AFRICA

TANZANIA
KENYA

ATLANTIC
OCEAN

INDIAN
OCEAN

N

THE EMPTY PLACE

NGORONGORO, SERENGETI AND NATRON LAKE

148
An African elephant browses for food in the forests of Ngorongoro. The largest living land mammals, elephants stand thirteen feet tall at the shoulder and weigh over fifteen thousand pounds. Their muscular trunks are used to feed themselves and to suck up water. An adult elephant consumes nearly five hundred pounds of forage each day, along with fifty gallons of water. Both males and females grow tusks of ivory, elongated incisors attached to the skull.

149 Top left
Hippopotamuses wallow in the water, with beautiful pink flamingoes gathered on the hillside behind them. Found only in Africa, Hippopotamuses can be traced through fossil evidence back to the Pliocene Epoch, nearly five million years ago.

149 Top right
The sweep of the Ngorongoro Crater's rim can be seen here, surrounding the beautiful, vast plain inhabited by so many fascinating creatures. The Serengeti-Mara-Ngorongoro ecosystem is one of the most important and least disturbed on earth.

Ironically, the Masai call it siringet, or the empty place. Its flat plains look empty at first sight, and few human beings live there. But it is the animals that rove this empty place that make it come alive. It is one of the last places on earth where people can see a plains wildlife habitat nearly undisturbed by man and unbounded by his fences and borders. The Serengeti-Mara-Ngorongoro ecosystem covers an area of over 18,000 square miles—Serengeti alone fills 5,700 square miles—and supports a diverse cast of wildebeests, lions, cheetahs, giraffes, elephants, and monkeys, totaling approximately more than three million creatures.

The Serengeti consists of vast areas of flat, open plains and others of rocky foothills. There are dry regions and patches of lush vegetation around rivers and streams. The southern and eastern areas are also dotted with kopjes, island-like, smooth rock formations that rise suddenly from the plains.

Fortunately, an early interest was shown on the part of African leaders to preserve wildlife. In Tanzania, a game reserve was set aside as early as 1905. Serengeti National Park and Research Institute was created in 1951 to preserve the nation's wildlife heritage, with boundaries forty miles long by fifty miles wide. As Tanzania achieved its independence from Great Britain in 1961, the "Arusha Manifesto" was issued in Tanganyika by J. K. Nyerere, the Prime Minister. It stated in part: "The survival of our wildlife is a matter of grave concern to all of us in Africa. These wild creatures amid the wild places they inhabit are not only important as a source of wonder and inspiration but are an integral part of our natural resources and of our future livelihood and well-being. In accepting the trusteeship of our wildlife we solemnly declare that we will do everything in our power to make sure that our children's grandchildren will be able to enjoy this rich and precious inheritance."

Today a unique combination of diverse habitats enables the Serengeti to support many species of large herbivores and carnivores and nearly five hundred species of birds.

150 Top left

The floor of Empakaai Crater can be seen in this aerial view, including its thickly forested walls and its 259-foot-deep lake. Virgin grasslands are cropped by African Cape buffalo that graze throughout the area. Incredible views of distant Serengeti, Ngorongoro, and Kilimanjaro can be seen from Empakaai.

150 Bottom left

Majestic King of the Beasts, an African lion rests on the Ngorongoro plain. Lions hunt large grazing animals in bursts of energy, eating as much as forty pounds of flesh and then relaxing for up to a week until their next kill. An adult male can grow up to eight feet long, plus a forty-one inch tail, and weigh 550 pounds.

150 Top right

A member of the horse family, the zebra is clad in brown and white stripes that break up its outline enough to throw off predators. Zebra graze alongside wildebeest and feed on the same grasses. Ngorongoro's zebras are the most numerous of three species of this African animal.

150–151

One of the African lion's favorite meals, wildebeest, also known as gnu, graze on the landscape of Ngorongoro. One of two gnu species in Africa, the blue wildebeest females form into herds with their young, while single males defend marked territory.

Ngorongoro

To the north of Serengeti, across the border from Tanzania, lies Kenya's Masai Mara Game Reserve. Over a million and a half wildebeest migrate northward each year into the Masai Mara in search of water and fresh grasses, then return to the Serengeti. The wildebeest is an ugly animal, with a huge bearded head, humped shoulder, and short hindquarters. As thousands of these animals plunge down the dry slopes and riverbanks to a ford, they never know when a crocodile might be hiding beneath the murky waters. Suddenly a hapless animal will feel vicelike jaws and knifelike teeth tear into its body, the closing mouth of the crocodile rising above the water with the squirming prey dying in its grasp. The muddy spray of the water flies off in all directions as the crocodile shakes its prey in a death grip. Thousands of animals flee the scene, moving onward.

Prides of lions sit lazily in the shade of copses of trees, until hunger motivates them to rise and hunt. These beautiful beasts move with their slinking gait to approach their prey among the huge herds of zebra and Thompson' gazelles. After choosing a potential victim, their charge is as determined and frighteningly aggressive as anything in the animal kingdom. The plain is shared with large numbers of other magnificent creatures, such as giraffes, hyenas, and cheetahs.

Elephants must eat several hundred pounds of leaves each day to retain their ideal weight of 14,000 pounds. They can often be seen crashing through the bush in the more wooded areas of the reserve.

Just to the east of Serengeti is the Ngorongoro crater, the world's largest perfect or unbroken caldera. Ngorongoro is over eleven miles across, with walls rising two thousand feet straight up from the crater floor. High mountain forest extends to the lip of the crater. Crotons dominate the low altitudes as one approaches the caldera, and blend into red

151 Bottom right
When asked which animal was the most formidable he had ever stalked, avid hunter Theodore Roosevelt—who had hunted lions, elephants, bison, grizzly bears, and rhinos—unhesitatingly named the African cape buffalo. Powerful and muscular, standing five and a half feet at the shoulder, the cape buffalo sports a dangerous rack of three-foot horns. Even lions are wary of a healthy cape buffalo.

151 Bottom left
Topi graze in Ngorongoro National Park. Related to the wildebeest and hartebeest, topi look less outlandish than do their more bizarre cousins. Topi have strong legs for fast escapes from predators, and are known for their ability to leap tall obstacles.

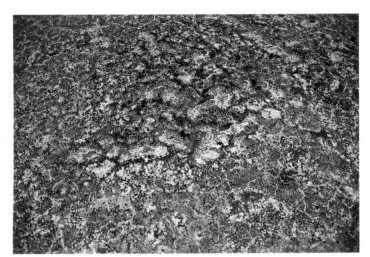

Pockmarked with thousands of ancient volcanic craters, the area of Lake Natron is accessible only by four-wheel-drive vehicles. Game is sparsely distributed across the frightfully hot desert pan.

Sodium carbonate salts leach into Lake Natron in this view. Sunset often tinges the algae-laced water pink, while the distant escarpment walls turn a brilliant red.

LAKE NATRON

seen from the height of a nearby peak, such as the active volcano Lengai. Looking down upon the dramatic rift escarpment below the peak, one's eyes move toward the beautiful blue lake to the east, surrounded by thousands of cones of soda volcanoes. The rough, pitted land, the colors in the rock and the calm blue tints of the water create an unforgettable sight, particularly at sunrise or sunset.

The lake lies in the midst of the land of the Masai people, a hot, unforgiving desert with little fresh water that is seldom visited by outsiders. But within this harsh realm lies incredible life. Walk along the walls of the escarpment growing with succulent euphorbias and dry scrub, and suddenly come upon an oasis with date palms, fig trees, and a waterfall. Walk along the flats of crystallized soda and look out upon the lake, the principal breeding ground for the Rift Valley's flamingoes. After a rain the escarpment walls blossom into the brief yellow flowers of the delonix tree. There is much beauty in this region, sometimes inhabited by an occasional wildebeest, zebra, or giraffe on the plain near the lake.

Altogether, the incredible landscapes of the Serengeti, Lake Natron, the Ngorongoro Crater, and their affiliated parks provide precious habitat for some of the last remaining wild roaming creatures of the earth, living in the great numbers and density of their prehistoric ancestors during the Pliocene and Pleistocene Epoch. The vision of these animals, silhouetted against classic African landscapes, forms one of nature's most magnificent wonders.

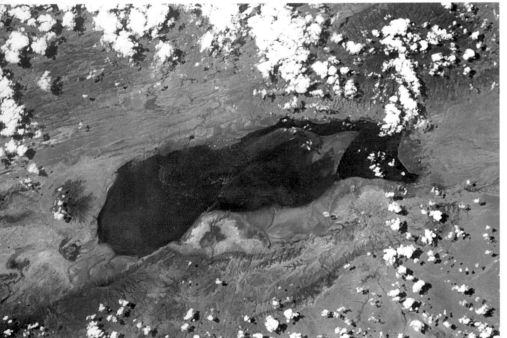

Lake Natron sits astride the thirty-five-mile-wide East African Rift Valley and is surrounded by sodium carbonate volcanoes. Salts from these volcanoes, like those seen in the foreground here, are transported into the lake by erosion.

The entire outline of Lake Natron can be seen in this view taken from space in 1993. Shades of red in the lake are due to the water chemistry, seen here at a peak of activity. White spots on the lake bed are drying soda salts. Blue water on the right indicates circulation and depth.

A strange landscape composed of dry salts and pitted earth surrounds Lake Natron. The beautiful blue walls of the rift escarpment can be seen in the background. Although the clear portions of the lake are blue, algae and alkaline salts turn the northern portion red.

Thousands of pink flamingoes take flight over Lake Natron, their principal breeding ground. While their nests are located in isolated areas deep within the salt flats in the central area of the lake, they often congregate in the blue, clear southern end to feed.

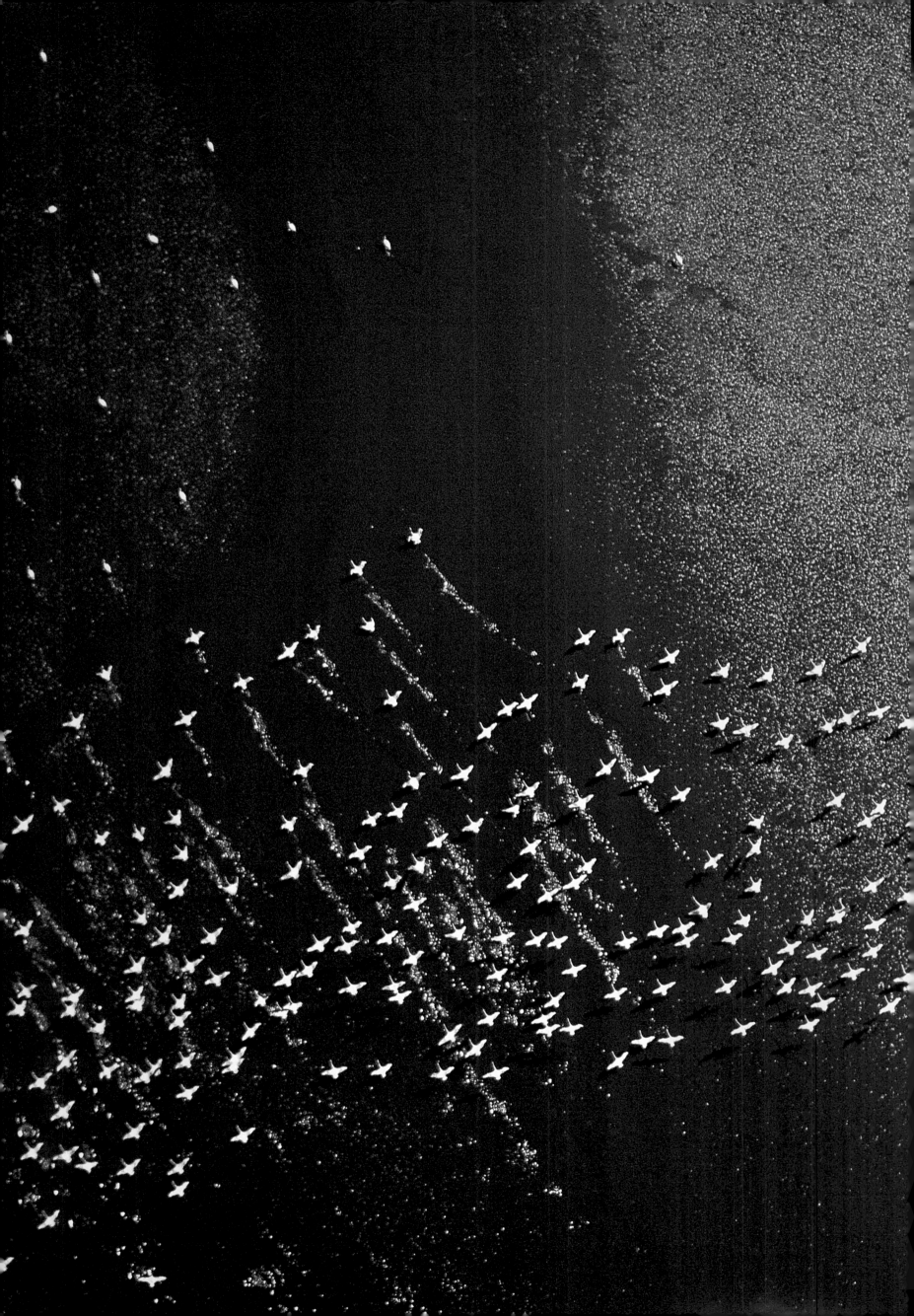

ATLANTIC OCEAN

AFRICA

N

TANZANIA →
UGANDA
KENYA

ATLANTIC OCEAN

INDIAN
OCEAN

THE HEART of AFRICA

LAKE VICTORIA AND RUBONDO ISLAND

The boat cruises slowly over the surface of the clear blue water. In the green reeds to the right, hippos are feeding. Suddenly, an aggressive male makes a lunging, false charge toward the boat, but soon he is left in the wake. Along the shore, Nile crocodiles bask in the sun, while the branches of the trees are full of birds, searching the waters for a meal. The lake is so large that it almost qualifies as an inland sea, and the distant horizons do little to dispel the notion. This is not an ocean or a sea, however, but a freshwater lake located deep in the heart of Africa. Lake Victoria, also known as Victoria Nyanza, is the largest lake in Africa and the main reservoir of the Nile. Three nations, Tanzania, Uganda and Kenya border it. Measuring 26,828 square miles, it is the second largest freshwater lake in the world after Lake Superior in North America. At an elevation of 3,718 feet above sea level, Victoria has a hot,

humid climate. Sugar cane is grown on the plains around the lake, and the local people fish in its waters for their food. The lake's waters fill a shallow depression in the center of the great plateau that stretches between the Western and Eastern rift valleys. In 1954 the Owen Falls Dam was completed. It turned Lake Victoria into an even larger body of water, one of the world's largest reservoirs, and raised the level of the water by three feet. The dam, located at Jinja in Uganda, provides hydroelectric power for this, one of the most populous areas of Africa. Lake Victoria contains many archipelagoes and numerous reefs, often hidden just below the surface of her clear waters. Cliffs rising three hundred feet back the southwestern shore of the lake, but on the west coast these change to papyrus and ambatch swamps that line the delta of the Kagera River. The lake's deeply indented northern coast is surprisingly flat and bare. Such a large body

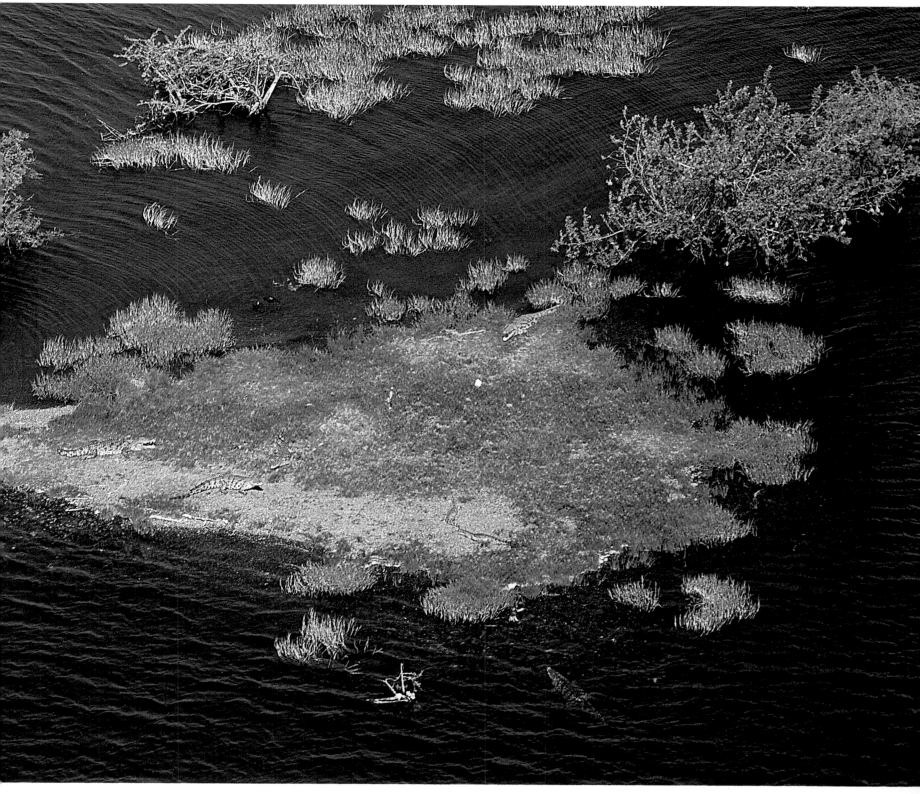

of fresh water in such a thirsty land is obviously the preferred habitat of many creatures, those who live above and those who live below the waves. The lake is home to more than two hundred species of fishes, dominated today by an introduced species, the Nile perch. These fish grow to weigh as much as two hundred pounds, and have greatly affected the native fish population, killing off up to 90 percent of the endemic species.

Along the shores of the lake and on its magnificent islands, scenery and wildlife are preserved. Ukerewe, a densely populated island located north of the Speke Gulf, is the largest of the lake's islands, with wooded hills rising 650 feet above the waters. At the lake's northwestern corner is the Sese archipelago, composed of sixty-two strikingly beautiful islands.

Rubondo Island is a Tanzanian National Park covering approximately 175 square miles in the southwest corner of Lake Victoria. Encompassing several types of vegetation ranging from savanna and open woodland to thick forest, Rubondo is noted for the Sitatunga, a species of antelope that inhabit its papyrus marshes. Chimpanzees, giraffes, elephants, and black rhinos have also been introduced to this gorgeous island sanctuary, whose woodlands come alive to the bright colors of butterflies, orchids, and red coral trees. The lush, thick greenery of the island provides food and shelter for many animals, including hippo, otter, bushbuck, and vervet monkeys. Nearly four hundred species of gorgeous birds have been documented on the island. The chief predator of the region is the crocodile with his fearsome teeth and jaws. In addition to a forest refuge, the island also includes tranquil beaches along the lakeshore and inviting nature trails. With no automobiles allowed, the quiet ambiance of the natural African ecosystem fills the island. Access to Rubondo Island is only by charter airline or boat. The wild, primitive nature of this sanctuary makes it well worth the journey.

Lake Victoria is a huge, beautiful refuge in the very heart of Africa. The lake is the place where the Nile rises, poised to course its way northward over half the continent. But Lake Victoria is also a special gem loaded with spectacular scenery and wildlife that qualify it as one of nature's wonders.

THE MOUNTAINS OF THE MOON

THE RUWENZORI AND VIRUNGA MOUNTAINS

You are hiking high above the African plains, through a forest of unearthly-looking plants. The mist is thick and visibility low. Suddenly the yellow disk of the sun appears above, burning through the moisture. The mist lifts and blue sky covers the area, while below a chain of alpine lakes can be seen, along with an incredible view of a flat plain and a snow-capped mountain chain. These are the Ruwenzori Mountains of equatorial Africa.

The Ruwenzoris are often referred to as the Mountains of the Moon. There are five individual peaks that support larger glaciers than any other on African mountains. The Ruwenzoris lie along the western border of Uganda and rise to a height of 16,763 feet, the measure of Mount Stanley, the tallest of the peaks and Africa's third-highest mountain. A World Heritage Site, Ruwenzori became a Ugandan National Park in 1991. Its seventy-mile length and thirty-mile width protects great visual splendor and important animal habitat.

Explorer Henry Morton Stanley was the first European to see the beautiful, mist-shrouded peaks, and he urged the use of the local name, Ruwenzori, or the hill of rain. Only thirty inches of rain per year fall on the plains below, while the Ruwenzoris are drenched with one hundred inches per year. The mountains are famous for their giant vegetation and moss-covered bogs, the result of so much precipitation.

The Ruwenzoris were formed when a giant block of the earth was left standing when the Western Rift Valley dropped down. Unlike most of the other famous peaks in Africa, the Ruwenzoris are not volcanic in origin. Clouds of mist almost perpetually veil the mountains from view, but when the sun breaks through and the mist dissipates, the mountains stand out boldly with their white caps against the blue sky.

The lush vegetation found above ten thousand feet includes giant forms of lobelia, heather, and groundsel, decades old and endemic in this form to this region of Africa. The mountains are home to blue monkeys, chimpanzees, hyrax, and the giant forest hog.

Across a valley from the Ruwenzoris, the Virunga Mountains form a volcanic range between Lake Edward and Lake Kivu, where the nations of Zaire, Uganda, and Rwanda meet. The Virungas are a part of the African lake district, on the western branch of the Great Rift Valley. In Rwanda the mountains are

162 Top left
The Virunga is a volcanic range between Lakes Edward and Kivu, near the point where the nations of Zaire, Uganda, and Rwanda meet. There are eight volcanoes in the Virunga. The highest, dormant Mount Karisimbi, rises to 14,787 feet.

162 Top right
A mother gorilla and her child are glimpsed amid heavy foliage. Mountain gorillas are the largest and rarest primate species. Extremely strong and intelligent, the average adult male stands at 5'6" and weighs 400 pounds, while the average female is 4'6" and weighs about 200 pounds.

162 Center
The lower slopes of the Ruwenzori, The Mountains of the Moon, are covered with dense forest like this. Eight primate species live in the region, including colobus, monkeys, chimpanzees, and baboons.

162 Bottom
A strange and beautiful wonderland opened up before explorer Henry Morton Stanley in 1889 as he explored the Ruwenzori Mountain Range. Fantastic plants, animals, and scenery stretched the limits of the known world.

163
A young mountain gorilla surveys the Virunga Mountains of Rwanda from a perch high in the trees. Mountain gorillas generally live in close-knit groups consisting of a silverback male (an older, dominant male with silver hair on its back), one or two subdominant males, several mature females, and their young.

164–165
The snowy face of Mount Ruwenzori presents a scene expected by few that visit the tropical continent. The mountain is topped by a permanent glacier and protrudes through several climatic zones.

164 Bottom left
Steam and fumes pour from the volcano at Nyiragongo, 11,369 feet high. The volcano features an active crater with a lava lake and has been highly active during the twentieth century, with major recent eruptions in 1977 and 1982.

164 Bottom right
This view was taken looking south from the summit of Mount Albert toward 16,763 feet high Mount Margherita and Mount Alexandra. Zaire is on the right, Uganda on the left. The gorgeous mountains of the Ruwenzori provide spectacular views of distant scenery as well as close views of exotic plant, animal, and bird species.

165 Left
This giant groundsel plant grows high in the Virungas at 12,800 feet. The area is located near Bujuku Lake and the north flank of Mount Baker can be seen in the distance. Other giant forms of plants, including lobelia and heather, also flourish in the area.

protected by the surrounding Parc National des Volcans and by Zaire's park of the same name, which stretches for more than two hundred miles along its border with Uganda.

The Virunga range contains eight volcanoes, some dormant, some active. The tallest, Mount Karisimbi, rises to 14,787 feet. Karisimbi, or white shell, is a huge, dormant dome with a white-capped summit created by hail and sleet. Muhabura, at 13,540 feet the second tallest, has a nearly perfect-looking cone. Gahinga has a flat top, Sabinyo is a jagged wonder, Visoke is enchanting with its slopes clad in green vegetation, while Mikeno is a spirelike apostrophe. The steep slopes are heavily forested and provide one of the few remaining habitats for the mountain gorilla.

A moist climate and rich volcanic soil have made the region below the mountains one of the most densely populated in Africa. The forests have been chopped back to the region of the volcanoes and the protection of the park.

Bamboo, that tree-size grass, dominates extensive areas of the region. The jungles of the Virunga tend to be cold, wet, dark and muddy. They receive over sixty-seven inches of rainfall per year. Tangled with thick vegetation, the green slopes of Visoke open out at altitude to a hagenia-hypericum forest, strong with the scent of dampness. The twisted trunks and huge, gnarled limbs of the hagenia trees are covered with dark green moss, ferns, and orchids, its green leaves covered with lichens. Wild flowers grow below these strange trees, and there is even St. Johns Wort, with delicate leaves and huge yellow flowers.

Above such forests is a subalpine zone, wet with rain year-round. Sun may dissolve quickly into swirling mist or rain in this brooding, beautifully gray environment. The climate here has allowed many plants to attain gigantic size. These include lobelias, noted for their huge vertical flower spikes. Also found are giant groundsels or senecios. Related to the sunflower, the groundsels grow to over twenty feet high in the Virunga and Ruwenzori, and can enjoy a life span of more ethan one hundred years. Their stems are thick and corklike with strange, smooth, stunted arms that hold large leaves. Yellow flowers grow on long spikes.

Most animals of the region are nocturnal, and are rarely seen by the casual tourist. The most famous are the mountain gorillas, studied by the late Dian Fossey from the 1960s until her murder in 1985. These huge, intelligent creatures are predominantly black, the males often having a telltale tinge of silver along their backs. They have highly developed societies and are fiercely secretive, reclusive creatures. Today, only about three hundred mountain gorillas survive in the wild and these live exclusively in the Virunga range. Widespread poaching and kidnapping of the young for zoos has eased off a bit as tourist dollars have begun to provide an alternative means of making money. As long as tourists are interested in seeing the gorillas in their habitat, the animals are worth more alive in the wild than dead. If these threatened animals can be secured and preserved in their natural surroundings it will be a victory for humankind.

Other animals that roam the mountains are elephants and buffalo, both dangerous when they feel threatened. There are also many unusual animals. The shy bushbuck is a small reddish-brown antelope with spiraled antlers that digs for roots, and browses through the trees. The screeching, horrid cry of the tree hyrax may be heard at dusk as it eats leaves in the bushes and small trees. They look like small rodents, but are more closely related to the elephant than any other creature. The rare golden monkey lives only in the Virunga Mountains in the bamboo zone. Giant earthworms may also be seen, some blue in color, growing up to a foot in length and much bigger in girth than an ordinary earthworm.

Altogether, the Ruwenzori mountains and the volcanoes of the Virunga present wildlife and vistas seen nowhere else in the world. Their mystery is summed up in the nickname, Mountains of the Moon, and the life forms that have evolved there are certainly unique. Yet it is also the familiar forms of elephants and gorillas, and their unique dependence on this environment, that lend such great beauty and significance to this beautiful region.

168–169
A panorama of the Namib Desert is spread out below the Space Shuttle in this view. The long, narrow expanse of the Namib, and the way it fronts on the Atlantic can be clearly seen. The Atlantic coast in this area is known as the Skeleton Coast since shipwrecked sailors faced almost certain death if stranded there. Today, a handful of people live in the area, primarily diamond miners and armed guards.

168 Bottom
Beautiful yellow flowers bloom each spring on the fringe of the red dunes at Sossusvlei, Namibia. This is a region usually devoid of such color, for its extreme aridity prevents the growth of many species. Atmospheric conditions and ocean currents in the South Atlantic prevent moisture from reaching the interior.

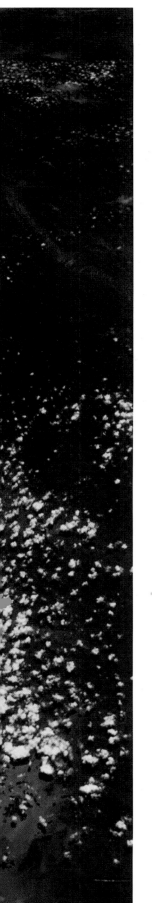

AFRICA

NAMIBIA

INDIAN OCEAN

ATLANTIC OCEAN

PLACE OF NO PEOPLE

THE NAMIB DESERT

169 Top
The vitaceae, a member of the grape family, looks almost as though it was uprooted and turned upside down. It is one of many plants and trees in the Namib region, like the baobab, which look more dead than alive to the uninformed viewer.

169 Center
The Namib is the oldest desert on earth. Here, mud cracks the dry plain in Namib-Naukduft Park,

one of the country's largest and most rugged conservation areas. The park includes a zebra reserve and incredible dwarf welwitschias, a strange prehistoric plant that lives about one thousand years.

169 Bottom
In addition to its famous sand dunes, the Namib also includes jagged mountain peaks and vast canyons, replete with weathered stones like these in Fish River Canyon.

From high in the air you see a vast blue ocean to the left, a vast yellow desert to the right. On one side, waves crest in white foam, on the other, dunes crest in brown shadow. Each world, that of the ocean and that of the desert, extends toward the horizon. Here is the essence of life, water, and the essence of lifelessness, an arid desert. These two worlds collide along a 1,180-mile strip of land on the southwestern coast of Africa known as the Namib Desert. Although from the air the ocean and the desert look like two entirely different worlds that turn their backs on one another, the cold, wet ocean and the hot, searing desert actually interact and join here.

The Namib is the oldest desert on earth, renowned for its wild and barren beauty. It is a sea of sand, whipped by the wind into plumes and whirlwinds. The name is taken from a Khoikhoi word that means desert or place of no people. The Namib is the only desert in the world where endemic flora and fauna have evolved in otherwise vegetationless dunes. The dunes are in effect totally isolated from other ecosystems in a desert that has existed for three to four million years. The desert first formed here as much as fifty-five million years ago, but the land was tempered about fourteen to eighteen million years ago to a semi-arid region. Hyperaridity returned with the advent of the cold Benguela current in the Atlantic Ocean about four million years ago.

The gigantic, sharp-crested red and gray dunes of the Namib are the highest in the world, averaging 525 feet tall. The highest dune, named Sossus Vlei, is more than 1,280 feet tall. Each dune can be as much as thirty miles long,

170–171
Beautiful long shadows are cast by the rising sun on the red dunes and green vegetation of the Namib at Sossusvlei. The gigantic dunes here are the world's highest at over a thousand feet. Many of the plants and animals are endemic to the Namib and depend upon mist from the Atlantic Ocean for survival.

170 Top left
The rising sun highlights the dunes of Sossusvlei, part of the harsh, primeval realm of the Namib Desert. The desert runs for 1,200 miles along the ocean in a narrow strip never over eighty miles wide.

170 Top right
Namib Desert dunes come in four different formations. Draa are composed of regularly shaped ripples one and a half feet tall on the slopes of larger dunes. Shrub-coppice dunes form on the downward side of bushes or grass. They are not tall but can extend up to a mile in length. Barchans are the crescent-shaped dunes found on the seashore. Mobile, they can move up to fifty feet each year. Star dunes are formed by winds coming from several different directions, and are found inland.

170 Bottom
A tree grows in the shadow of a huge dune in the Namib. About fifty-five million years ago, an arid climate began in the region, and a "sand sea" was formed up to 720 feet deep, more extensive than the current desert. The Tsondab Sandstone Formation is the remnant of this ancient desert.

and is spaced about a mile from its neighbor dune. The colors of the dunes change according to locality and age, the newest sand particles being near the ocean on the west. The coastal dunes are a light yellow-brown, while inland dunes tend to be a deep brown. In the eastern portion of the desert, furthest from the Atlantic, the dunes turn to a brick red color. The "sand sea" of the region covers 13,125 square miles. Although the desert is quite long, it averages only sixty to eighty miles wide between the ocean and the mountains to the east.

171 Right
Today, the Namib's "sand sea" covers an area of 13,125 square miles and consists of sharp-crested dunes oriented from south to north in lines. Sand grains are composed of 90 percent quartz, with small amounts of garnet, feldspar, and monazite making up the rest.

Many of the plants and animals of the Namib are endemic to this desert and depend on the clammy mists and fog that roll in off the Atlantic Ocean for their survival. In addition to water, the bountiful Atlantic Ocean also provides other means of supporting life. The winds blow surface ocean waters northward, allowing cold water to surface, bringing rich nutrients with it. Fish feed on these nutrients, thus passing them from organism to organism as one form of life is consumed by another, up and up on the food chain. Since there are few watering holes, animals must roam the beaches and seek nourishment from the bountiful marine life.

Although the sands seem barren, more than two hundred species of beetles, scorpions, spiders, geckoes, chameleons, snakes, and eagles have adapted to survive there. Splashes of greenery along the coastal plain reveal old lichens, while a strange remnant of prehistoric flora, the welwitschia, lives here as well. A dwarf tree that survives on the moisture of the Atlantic coast, its life-span is estimated at more than one thousand years. Tall stalks

of aloe grow on the fringes of the desert, but little other than lichen grows in the desert itself.

At dawn each day the cycle of life in the Namib begins. As the sun rises slowly and colors the sky a watermelon pink, warm air from the Atlantic Ocean sweeps over the cold waters of the Benguela current. The contrast in temperatures produces a thick coastal fog that can penetrate as much as sixty miles inland. The little drops of condensed water from these fogs sustain many plants and animals from day to day. Beetles stand on their hind legs to catch the moisture, sidewinder snakes lick it from their own bodies, and dune ants drink it off blades of brown grass. The rising sun gradually evaporates the remaining moisture in the air, and the Namib becomes a world of golden, rolling sand dunes. Ostriches, vultures, warthogs, oryx, and sand grouse gather at muddy water holes to drink and prepare for the hot day ahead.

The day heats up quickly and by afternoon the sun

174–175 and 174 bottom
The easternmost of the Seychelles islands, La Digue is about two miles long and rises to 1,071 feet in the center. Gorgeous beaches and date palms, as well as naturally sculpted rocks, line the ocean shores.

175 Top
The coral atoll of Aldabra, located northwest of Madagascar, is seen in this aerial view. As much as 10 percent of the plant and animal life of Aldabra is endemic to the atoll. The inner lagoon and the branchlike deep blue channel, which allows an exchange of seawater, is clearly visible here.

and is spaced about a mile from its neighbor dune. The colors of the dunes change according to locality and age, the newest sand particles being near the ocean on the west. The coastal dunes are a light yellow-brown, while inland dunes tend to be a deep brown. In the eastern portion of the desert, furthest from the Atlantic, the dunes turn to a brick red color. The "sand sea" of the region covers 13,125 square miles. Although the desert is quite long, it averages only sixty to eighty miles wide between the ocean and the mountains to the east.

Many of the plants and animals of the Namib are endemic to this desert and depend on the clammy mists and fog that roll in off the Atlantic Ocean for their survival. In addition to water, the bountiful Atlantic Ocean also provides other means of supporting life. The winds blow surface ocean waters northward, allowing cold water to surface, bringing rich nutrients with it. Fish feed on these nutrients, thus passing them from organism to organism as one form of life is consumed by another, up and up on the food chain. Since there are few watering holes, animals must roam the beaches and seek nourishment from the bountiful marine life.

Although the sands seem barren, more than two hundred species of beetles, scorpions, spiders, geckoes, chameleons, snakes, and eagles have adapted to survive there. Splashes of greenery along the coastal plain reveal old lichens, while a strange remnant of prehistoric flora, the welwitschia, lives here as well. A dwarf tree that survives on the moisture of the Atlantic coast, its life-span is estimated at more than one thousand years. Tall stalks

of aloe grow on the fringes of the desert, but little other than lichen grows in the desert itself.

At dawn each day the cycle of life in the Namib begins. As the sun rises slowly and colors the sky a watermelon pink, warm air from the Atlantic Ocean sweeps over the cold waters of the Benguela current. The contrast in temperatures produces a thick coastal fog that can penetrate as much as sixty miles inland. The little drops of condensed water from these fogs sustain many plants and animals from day to day. Beetles stand on their hind legs to catch the moisture, sidewinder snakes lick it from their own bodies, and dune ants drink it off blades of brown grass. The rising sun gradually evaporates the remaining moisture in the air, and the Namib becomes a world of golden, rolling sand dunes. Ostriches, vultures, warthogs, oryx, and sand grouse gather at muddy water holes to drink and prepare for the hot day ahead.

The day heats up quickly and by afternoon the sun

171 Right
Today, the Namib's "sand sea" covers an area of 13,125 square miles and consists of sharp-crested dunes oriented from south to north in lines. Sand grains are composed of 90 percent quartz, with small amounts of garnet, feldspar, and monazite making up the rest.

172–173

Four Oryx use their energy to climb a dune in the Sussusvlei in Namibia. The Namib Desert can be an extremely inhospitable place, yet animals of all sizes, different forms of plant life, and birds have all found ways to survive there.

scorches the desert sands, sending animals to find cover from its remorseless rays. Along the sandy beaches, fur seals loll within range of the ocean's spray, wary of predators like jackals and hyenas. The seals come to these shores each autumn to give birth and have to be particularly careful to guard their newborn pups. Seabirds like flamingos, pelicans, and terns fish in the shallow offshore waters, while endangered jackass penguins, endemic to this area, play in the surf.

Finally, the long hot day comes to an end, and the animals that depend upon sunlight for survival, especially reptiles, sleep. The yellow moon rises above the cooling sand as owls, spiders, scorpions and tiny, nearly transparent geckos with their loud mating calls emerge and search for food before the next sunrise.

This is the endless cycle of life in the Namib. Specialized creatures have adapted to life in these towering sand dunes, their gorgeous undulating waves making it easy to forget just how difficult it must be to eke out an existence there. But animals and plants adapt and survive. In this incredible wonder of nature, this place of no people, even here there is life.

173 Top
A flock of ostriches runs across the Etosha Pan. Feast or famine is the motto of the Etosha, an immense depression in Namibia that dazzles with its heat waves and mirages. At certain times of the year, the pan floods and is covered with water, providing life for thousands of animals. After a few weeks the sun burns off the water, leaving behind salts.

173 Center
This oryx did not survive the heat, lack of water, or sparse food supply in the Namib. The desert is not the preferred environment of the oryx, although they frequently cross the dunes. Oryx can survive without water if they find sufficient wild herbs and melons.

THE NAMIB DESERT

173 Bottom
For a land so thirsty, portions of the Namib are able to support a wide range of larger animal species. In this view, a small herd of elephants heads toward water in the Hoanib River basin in Kaokoland. Namibian rivers rush with water for only a few hours each year, but replenish precious underground water resources that last all year long.

174–175 and 174 bottom
The easternmost of the Seychelles islands, La Digue is about two miles long and rises to 1,071 feet in the center. Gorgeous beaches and date palms, as well as naturally sculpted rocks, line the ocean shores.

175 Top
The coral atoll of Aldabra, located northwest of Madagascar, is seen in this aerial view. As much as 10 percent of the plant and animal life of Aldabra is endemic to the atoll. The inner lagoon and the branchlike deep blue channel, which allows an exchange of seawater, is clearly visible here.

175 Bottom

Aldebra's main channel, which allows the exchange of ocean water with water in its lagoon, can be seen here. Aldebra is composed of thirteen islands, and has two distinct terraces of life zones, at twelve feet and twenty-four feet above sea level.

AFRICA

ATLANTIC OCEAN

SEYCHELLES

INDIAN OCEAN

INCOMPARABLE BEAUTY

THE SEYCHELLES ISLANDS

The slanting trunks of palm trees stretch over perfect, pristine sand beaches. The sunlight filters through the palm fronds, hitting the greens and oranges of the leaves. Wooded trails lead through cashew thickets, bamboo glades, and citrus and avocado plantations. A pristineblue ocean breaks over pink rocks, foaming white on the shore. These are islands of incomparable beauty, teeming with rare and unusual creatures. These are the Seychelles.

The Seychelles have been called Paradise on Earth. Not as famous as the islands of the South Pacific or the Caribbean, they have every bit as much of their beauty. Located in the west central Indian Ocean, a thousand miles from the east African coast, this archipelago of about 115 islands stretches for hundreds of miles. The Seychelles are located between 4 and 5 degrees south of the equator, and have a consequently warm, wet climate. The three principal islands—Mahe, Praslin, and La Digue—are granite, while the outlying islands are coral atolls. Surprisingly, the Seychelles are not volcanic in origin. They are the peaks of a huge underwater continental plateau that separated from India about sixty-five million years ago.

The granite islands support lush forests on their mountain slopes. The coral atolls are also dense with vegetation, including palm trees, casuarinas, banyans, and tortoise trees. The giant coco de mer palm is endemic to the islands, and bears a strange fruit that looks like the lower torso of a nude woman. These trees can live up to one thousand years. The islands have few species of mammals, but are rich in insects, tropical fish and birds.

The main island of Mahe is known for its idyllic beaches and plantations of coconut palms and cinnamon. Huge, incredibly shaped granite boulders dot Praslin, while offshore coral reefs teem with over nine hundred species of fish. On the north side of the island, a beach astounds visitors with its glorious white sand and its large, rounded granite boulders. Stunning La Digue is probably the most-photographed island in the world, with several rare species of birds. The rocky shoreline, topped with palm trees, seems to tumble into the sea.

Bird watchers descend on the Seychelles to see some of the rarest birds on earth. Eleven species are endemic to the islands, including the magpie robin, the Seychelles warbler, and the only remaining flightless bird in the Indian Ocean region, the white-throated rail. Migrant birds pass through from the Arctic and Antarctic. Cousin Island is owned by the Royal Society for Nature Conservation, and it allows small groups to visit its vibrant bird habitat on Tuesdays, Thursdays, and Fridays.

With five times as many giant land tortoises as the Galapagos, the island of Aldabra is a World Heritage Site. This is the original habitat of the tortoises, and about 200,000 live there. Aldabra also harbors water birds like egrets, sacred ibis, terns, and flamingoes. The locals call strange mushroom-shaped pinnacles of limestone, which are honeycombed and pockmarked, champignons. The island is one of the world's largest coral atolls, fourteen miles in diameter. It encloses a huge tidal lagoon, sometimes home to tiger sharks and manta rays as well as millions of colorful tropical fish.

The Seychelles are some of the most beautiful islands on earth, abundant in nature's bounties, including silver white sands, unbelievable coral reefs, and unique plants and animals. Like the continent of Africa, they are mysterious, wholly unpredictable, and fascinating in their details.

176–177
The red sandstone formations of
Monument Valley encompass
about two thousand square miles
in northeastern Arizona and
southeastern Utah. Tall red
sandstone rock formations rise as
high as a thousand feet from the
surrounding plain.

176 Bottom left
Amazing cactus giants are seen in
Saguaro National Park located
near Tucson, Arizona. This desert
haven preserves these
endangered, slow-growing cacti
along with undulating desert hill
and scrub country.

176 Bottom right
The gorgeous "red rock" country
of west central Arizona near
Sedona is watered by Oak Creek
Canyon and dominated by
features like Cathedral Rock. This
large area of red sandstone

formations yields to dramatic
hills and mountains just north of
this site, with vegetation and
terrain shifting from desert
plants to mountain conifers in a
matter of a few upward-climbing
miles.

NORTH AMERICA

INTRODUCTION

Flocks of birds rise from frigid northern lakes to wing their way southward over arctic tundra, spectacular mountains, flat fertile river valleys, and the deserts and jungles of Mexico. In flight they pass over waving fields of grain, rough snow-capped peaks, and the slow meandering rivers that carry and deposit fertile soil into the river valleys. Nature has blessed this continent with vast agricultural potential, good soil, and abundant water. Extraordinary places are protected and revered here, including deep red rock canyons, sculpted pinnacles and arches in the dry deserts, spectacular waterfalls, and incredible peaks.

This is North America, the world's third largest continent It is shaped roughly like a large triangle, with the upper western point nearly touching Asia in the Bering Sea, the upper eastern point looking toward Europe, and the southern point connecting it with Central and South America.

North America comprises 9.4 million square miles and is 4,500 miles long from the Canadian Arctic to the border of Panama and 4,000 miles wide across Canada. It has many great harbors and ports, particularly on the craggy Atlantic coast. Although there are many rural areas, the country is becoming more and more urbanized every year. It is rich in such natural resources as minerals and agricultural products.

There are seven major land regions in North America, beginning with the Pacific coastland, which stretches from Alaska to Mexico down the west coast of the continent. It consists of beautiful, rocky coastlines with sandy beaches amid pinnacles on the shores, and ranges from the wet rainforests of the north to the dry desertlike conditions of southern California and Mexico. Throughout this region run incredibly beautiful mountains, including the coastal range just inland from the Pacific and the continent's inner ranges—the Alaska Range, the Cascades, the Sierra Nevada, and the Sierra Madre. These high mountains include wonderful scenery, such as Mount McKinley in Alaska, the continent's highest point, and the Yosemite Valley in California, an extreme and superb example of a mountain ecosystem at work.

177 Top
Waves pound the shores of the California coast, creating amazing rock formations and beautiful sandy beaches. From the border of Oregon down to Mexico, the coast varies in condition and population, but is surprisingly natural and pristine, with little development and large protected areas administered by the California State Park System and the National Park Service.

177 Bottom
South of Carmel, California, is a beautiful stretch of coastline, popularly known as the Big Sur Country. There are few towns along the highway, only scenic views and twisting, winding roads that follow the contours of the coast and canyons. A drive along the coast is spectacular, with a seemingly endless abundance of surprising vistas.

178–179
A dusting of snow heralds winter in this view of Banff National Park, in Alberta, Canada. The park was created in 1885 and contains 2,564 square miles of incomparable Rocky Mountain scenery. Banff adjoins Jasper, Kootenay, and Yoho National Parks and three provincial parks that form the Canadian Rocky Mountain Parks World Heritage Site.

178 Top left
The massive, snowy expanses of western Canada encompass an enormous region of the northern Rocky Mountains.

178 Top right
Mount Rainier is a currently dormant 14,410-foot volcano in the State of Washington. Mount St. Helens, located only thirty miles to the southwest, erupted in 1980. The West Coast of the United States lies along several fault zones, making it far more subject to dramatic volcanic and seismic disturbances than other parts of the country.

178 Bottom
Baja California's beaches offer mile after mile of scenic splendor and recreational opportunities. Sandy stretches of shore are popular for surfing, deep-sea fishing, swimming, sailing, and scuba diving. The Pacific Coast's wonderful sandy beaches are pleasant year round and extend from southern California into Mexico.

The second major geographical area of North America is the Basin Region, the lands between the coastal mountains and the Rocky Mountain chain. This region includes the Yukon River basin in Alaska, the interior plateau of British Columbia, the Colorado Plateau, the Great Basin, and the Mexican plateau. The area includes some of the continent's most spectacular works of nature, such as Grand Canyon of the Colorado River, which runs a mile below its red rock rims, and Monument Valley, with fascinating carved pinnacles that show the incredible results of erosion. The region also includes Death Valley, which is the lowest point on the continent.

Next come the magnificent Rocky Mountains, the high inland spine of North America. Awe-inspiring snow-covered peaks rise above plains on either side. Canadian parks such as Jasper and Banff and American parks like Rocky Mountain and Yellowstone have been created to preserve the most gorgeous and spectacular areas of this alpine wonderland.

The Great Plains cover an enormous area in the midwestern United States and central Canada, with the Rockies to the west and the Appalachian Mountains to the east. In between lies a vast rolling plain dominated by the Mississippi River system. The river winds for mile after mile through the region, absorbing other rivers and fertilizing this breadbasket of the Americas.

The Canadian Shield covers the eastern half of Canada and the northeastern United States. It is called a shield because it is composed of ancient rock that has not been geologically active for hundreds of millions of years. The mighty Great Lakes lie upon the shield. They are so large that they look like oceans, creating their own sand dunes and at times showing a moody, stormy side. The roaring cataract of Niagara Falls rushes over a fault in the Canadian Shield, demonstrating nature's power and exuberance.

The Appalachian Mountains run parallel to the Atlantic coast, and never exceed 6,700 feet at their highest point. These are very old mountains, worn down by eons of wind and weather.

The Atlantic Coastal Plain is the seventh and last of North America's geographical regions. It is a lowland area that extends from Cape Cod in Massachusetts around the tip of Florida, then along the Gulf of Mexico to the Yucatan Peninsula. It is characterized by sandy beaches with dune systems and barrier islands.d It is sometimes a sunshiny paradise, at other times a storm-lashed, unprotected flatland ruled by the sea.

For hundreds of years this fertile, temperate continent has drawn settlers from all over the world. It possesses a huge number of nature's incredible wonders, on display for all to see and enjoy.

179 Right
Every fall the trees in the eastern United States in the Appalachian Mountain range begin to turn color as the leaves die. The many species of deciduous trees that inhabit the region have leaves with different pigment structures that turn an amazing variety of brilliant colors. This leads to wonderful autumnal views like this one near Woodstock, New York.

THE TOP
OF THE
WORLD

THE CANADIAN ARCTIC AND GREENLAND

180
The ghostly aura of the aurora borealis is seen here over the Canadian Arctic. The phenomenon constantly takes on different looks. Also known as the Northern Lights, the aurora borealis provides a spectacular show during the long Arctic nights. Different colors are caused by ionized nitrogen and oxygen in varying amounts. Only 3 to 4 percent of the energy can be seen by the human eye.

180–181
The surprisingly rocky and craggy northern Canadian topography in this view showcases the scenic beauty of this part of the earth. During the summer months, this strait in Lancaster Sound is free of snow and ice, allowing ships to pass from Baffin Bay through the Northwest Passage.

The orange glow of the sun, low in the sky, silhouettes icebergs floating in the frigid waters of the Arctic Ocean. Their blue-white shapes cut across the brightly lit horizon as flat mesas, pyramids, and pinnacles. Waterfall streams flow over precipices to the sea, where the tail of a humpback whale breaks the water, unbelievably large, black, and slick. Dramatic glaciers move off a seemingly endless ice cap in the clear, crisp Arctic air. This is the world of the far north—barren, uncompromising, yet incredibly beautiful.

The region is composed of a series of islands north of 70 degrees north latitude that include over half the island of Greenland and several islands that are part of Canada's Northwest Territories. The Canadian islands include Banks Island, Victoria, Prince of Wales, The Queen Elizabeth Islands, Somerset, Devon, Axel Heiberg, Ellesmere, and Baffin islands. These are located in the Arctic Ocean, the Beaufort Sea, and Baffin Bay. The islands are covered with permanent ice caps, and around their northern shores lies the Polar ice pack.

Many of the Canadian islands have a low, barren surface, but some in the northeast have mountains and icecaps that rise from six thousand to ten thousand feet above sea level. The cold climate and permafrost prevent trees from growing on any of the islands. A few people live in this area in scattered settlements and outposts. Native Inuit hunt, fish, and trade much as they always have, while other hardy souls run trading posts and live in small towns. Military bases for both the Canadian and United States armed forces are located here. Both maintain research stations and lookout posts in this barren country.

Greenland is the largest island in the world, covering an area of 840,000 square miles. Of this immense area, 708,000 square miles is covered with a permanent ice cap from one to two miles thick. The island is 1,660 miles long and 750 miles wide, and the highest point is 12,139-foot Mount Gunnbjorn, located east of the icecap. Greenland is a province of Denmark with home rule. The entire province harbors a small population of about fifty-five thousand people in a few sporadic towns clustered along the shoreline away from the ice, most below the Arctic Circle.

Greenland is a very old place, in fact the site of some of the oldest rocks ever dated, to 3,700,000,000 years ago. It is believed that Greenland's subsoil is at least 1,600,000,000 years old. Greenland is caught in the natural cooling effects caused by the cold ocean currents that surround the island and the ice cap that covers an area fourteen times the size of

181 Bottom left
The aptly named Elephant's
Foot Glacier is a feature of
Greenland National Park in the
northeastern part of the island

nation. Here the glacier can
be seen spilling through the
valleys between mountains from
the interior ice sheet of the
island.

181 Bottom right
The sun rises over raised
pressure ridges in Youngsund
Fjord in the northeastern
portion of Greenland National
Park. To protect the natural
resources of the island, about a

third of Greenland was
established as a National Park
in 1974. The Greenland park
is the largest national park in
the world, about the size of
France and England put
together.

England. The bottom layers of the ice are up to two million years old. During that time, the weight of the ice has pressed the original bedrock down about two thousand feet.

The ice cap can be visited from most points in Greenland, but must be done by boat or helicopter, for the province has no roads between towns or railroads. However, the ice cap is only about sixteen miles from Kangerlussuaq, from which a hike, a drive, or a bicycle trip can take you there. The huge ice formations take on the shape of arches, pinnacles, and tunnels in every conceivable shade of white, blue, and even orange when the sun rides low on the horizon. Gray glaciers slowly drain ice from the top of the baby blue ice cap, moving toward the ocean. The world's most active glacier, at Ilulissat, moves twenty-five yards a day and carves across a front over seven miles wide.

Towering white icebergs break off from the glaciers at the ocean, floating away from the coast and standing tall and white in the sun. In Greenland's Disko Bay, icebergs often rise up to three hundred feet above the waterline, giving an indication of their massive size, for at least 90 percent of an iceberg lies hidden below the waterline. Inevitably the icebergs drift out of the Arctic regions and gradually melt, a process that may take three or four years for the larger ones.

The region has an Arctic climate although there are great differences from north to south and between the coastal and inland areas. The climate is dry and temperatures seem warmer than they really are. During the Arctic summer the sun never sets. This what is called The Land of the Midnight Sun, where the sun does not set from late May to late July. During official summer "night-time" the soft warm light and long shadows from the sun, low to the horizon, make the land look dreamlike with their red, orange, and yellow tones.

In the winter, the sun does not rise above the horizon for several weeks. The landscape is white from the snow and the frozen sea. The only light comes from the stars, the moon, and the aurora borealis, or Northern Lights, creating an altogether different but still dreamlike study in blues, purples, greens, and blacks.

Autumn, winter, and spring bring the Northern Lights, which sweep across the dark sky above the snow-covered mountains. The Northern Lights come in colors of white, yellow, green and red, shimmering in the darkness like a curtain blowing in the wind. They appear all year round in these regions but can only be observed against a dark night sky. The Northern Lights are at a height of about sixty-five miles and are phenomena caused by electrically charged particles from the sun entering the earth's atmosphere and

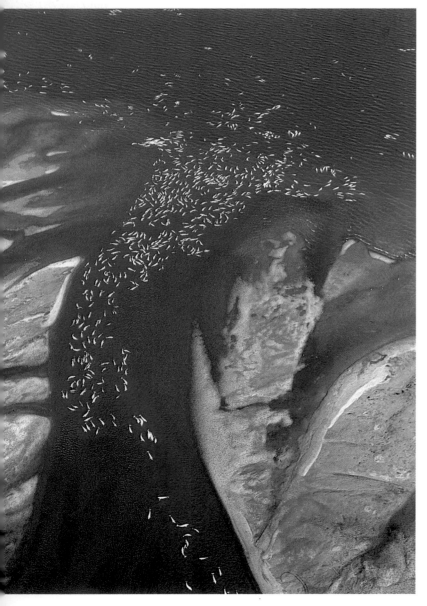

183 Bottom right
A polar bear jumps from one ice floe to another in the Arctic. Polar bears are among the largest terrestrial carnivores in the world, with some specimens weighing more than eighteen hundred pounds.

183 Bottom left
Two walruses are at home in the frigid waters of Hudson Bay. Enormous creatures, walruses grow up to eleven feet long and weigh from 1,800 to 3,700 pounds, with tusks as long as three feet. They bellow loudly and live in herds of several thousand. An entire herd will come to the defense of an endangered individual. Mollusks and shellfish are their principal diet, dredged up from the ocean floor.

188–189
The snow-covered surface of
Mount McKinley, North America's
tallest peak, looks quite placid in
this view. McKinley is instead a
cold, inhospitable place hidden
by clouds some 75 percent of the
time. Temperatures above 14,000
feet are severe even in summer,
and in winter can plummet to -95
degrees Fahrenheit. Winds can
exceed 150 miles per hour.

188 Bottom left
The West Fork Glacier flows from
the bottom right of this picture
toward the valley of the west fork
of the Chulitna River, seen
slashing across the center. The
Alaska Range began to form
sixty-five million years ago

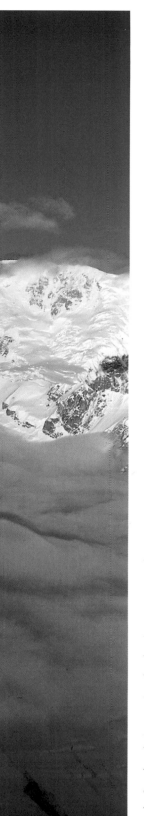

TUNDRA PARADISE

MOUNT MCKINLEY AND DENALI NATIONAL PARK

The sky is painted an icy blue as raw winds course between the craggy mountain peaks. A mantle of pure white snow glistens and sparkles on the face of the mysterious massif, with flecks and streaks of the gray rock peering out in serrated patches. The mountain looms above a snowy wonderland in winter that is transformed into a fragile tundra paradise in summer. Mosquitoes bother the flanks of a moose on a gray summer day as he wades through the shallow waters beneath the ever-white mountain named McKinley.

The Aleuts called it Denali, or the tall one. In 1896 settlers from the United States named it Mount McKinley after President-elect William McKinley. The mountain's southern summit rises to 20,320 feet, making it the tallest mountain in North America. The upper two thirds of the peak are covered with permanent snowfields that feed several glaciers, and in fact the massif stands seventeen thousand feet above the treeline. There is no life on its barren granite slopes, which are covered with ice hundreds of feet thick in many places.

The rugged sides of the perpetually snow-covered peaks of Mount McKinley rise above the line of the Alaska Range. They seem to ripple with the lines of snow and ice, as well as the faults and crevasses of the rock. The mountain seems to dominate all it surveys, an omnipotent force that can be seen for hundreds of miles around.

Because it is often shrouded in clouds, Denali for long has been a source of folklore and mystery. Climbers have tried to scale it unsuccessfully, beginning in 1903. The north peak was first scaled in 1910, and the taller south peak on June 13, 1913 by the explorer Hudson Stack and his three companions.

Mount McKinley is part of the arc-shaped Alaska Mountain Range, which stretches for six hundred miles from the Aleutian Range in south central Alaska to the Yukon boundary in southern Alaska. The range separates the interior tundra prairie of the region from the Pacific Coast. Mount Hunter, Mount Hayes, and Mount Foraker are three other peaks in the Alaska Range that exceed thirteen thousand feet. The Alaska Range was created by the movement of plates along the Denali fault, one of North America's largest breaks in the earth's crust. The mountain towers over all that surrounds it, and even creates its own weather. Powerful, sudden storms can pound the mountain and drive temperatures down.

Below the towering, craggy mountain is Denali National Park and Preserve. The park encompasses six million wilderness acres of tundra and taiga and protects a complete Arctic ecosystem. The area was originally established as Mount McKinley

188 Bottom right
A permanent mantle of snow embraces at least 50 percent of Mount McKinley even during the summer months. The core of the mountain is composed of granite and slate, covered by ice hundreds of feet thick in some places.

189
A craggy, unnamed peak stands beside the Ruth Gorge in the Alaska Range, Denali National Park. The gorge carries the Ruth Glacier down from the Alaska peaks to the park's rivers far below.

190–191
Snow and ice blankets the Savage River in Denali National Park. The Savage River flows to the north from the Alaska Range and runs close to the park headquarters at Riley Creek near the entrance.

190 Bottom left
The awesome forces of nature are on display at Denali. Here Mount McKinley is seen in the twilight. Soaring sixteen thousand feet above the surrounding landscape, McKinley stands apart from other mountains in the range. It is the tallest mountain in North America.

MOUNT McKINLEY
AND DENALI NATIONAL PARK

190 Bottom right
In 1980 the area of Denali was increased to six million acres of wild and unspoiled frontier. Mount McKinley towers over this wilderness paradise, but the park itself has so much more than just the mountain—fragile tundra ecosystems, fascinating wildlife, huge glaciers, and beautiful lakes.

191 Tcp
Light snow falls onto the tundra and summer flowers of Denali. Over 430 species of flowering plants, mosses, lichens, algae, and fungi grow in miniaturized *form in the thin layer of topsoil that thaws each summer. Sixteen to twenty hours of daylight help the plants to grow, mature, and die in a rapid life cycle repeated each summer.*

National Park in 1917, but the name was changed in 1980 to honor the original Aleut inhabitants. The park became an International Biosphere Reserve in 1976.

Taiga is a Russian term for a coniferous forest located in lower river valleys, and this portion of Denali lies at about twenty-seven thousand feet. The forest includes small spruce, aspens, balsam poplar, and birch. Once the elevation rises above three thousand feet the treeline is reached, above which only tundra, a gray-green ecosystem of small shrubs and tiny wildflowers, exists. The tundra has a short, cool, three-month growing season. Some of the tundra is dry and brittle at the higher elevations, while other tundra areas are cool and moist, with swampy fields of tussocks. Tundra actually comprises a miniature forest, with tiny versions of plants that can grow much taller and longer in other areas.

A small number of mammal and bird species have adapted to survival in the Arctic region, using layers of fat and fur to insulate them from the cold. Very large species of mammals have been successful here. The park provides a habitat for

191 Bottom
Evening light colors Ruth Gorge south of Mount McKinley. The Auletian Range runs along a fault *1,300 miles long from the Canadian border to the Alaska Peninsula. The region's tectonic action causes earthquakes and active volcanoes.*

192 Top
A grizzly bear mother and a rare brood of three cubs search for food in the summer tundra. Grizzlies are omnivores, eating small plants, berries, ground squirrels, moose, caribou calves, and even carrion if necessary. Like all bears, the sows are fiercely protective of their cubs and take violent, aggressive measures against any perceived threat.

MOUNT McKINLEY AND DENALI NATIONAL PARK

192 Center
The moose is the largest member of the deer family. Usually solitary creatures, not comfortable in herds, moose are native to both Europe and North America. The moose reaches its greatest size in Alaska, where a full-grown male may stand seven and a half feet at the shoulder and weigh up to 1,820 pounds. Antlers are shed each year after mating season. They grow back again each summer, reaching an adult spread of up to five feet.

192 Bottom
A female moose wallows in the shallows of a lake in the Alaska summer. Denali was original'y founded in 1917 to preserve the wildlife that surrounds Mount McKinley.

caribou, Dall sheep, moose, and grizzly bears. Large herds of caribou migrate to parts of the park, foraging for grasses and lichens. The solitary moose like to spend time in water in the lower reaches of the park, while Dall sheep prefer the higher elevations. The huge, unpredictable grizzlies have little fear of humans and can be extremely dangerous. Males weigh at least 450 pounds, stand well over seven feet tall when standing on their hind legs, and are characterized by the hump on their backs and their broad, flat faces. Grizzlies are the true kings of the North American wilderness.

The Denali region contains huge mountain ranges and flatlands stretching as far as the eye can see. It receives sixteen to twenty hours of daylight in the short summer season, while the winters provide very few hours of sunlight. Plants have to withstand prolonged low temperatures, then flourish and reproduce during the brief summer season. Regional plants are frost-resistant and low growing so that they can be sheltered from the wind by the winter snows.

Altogether, Denali impresses as much with its incredible, tenacious flora and fauna as with the massive, gorgeous granite peak. On a summer day, a distant view of the peak seems to show three horizontal bands—the lowest a lush green taiga forest, the middle the gray and white massif surrounded by misty clouds, and the upper a clear blue sky overarching all. The incredible beauty of Denali's entire ecosystem makes it one of nature's special wonders.

ALPINE WONDERS

JASPER AND BANFF NATIONAL PARKS, CANADA

194
Athabaska Falls in Jasper National Park is a roaring wonder, a particularly spectacular waterfall, one of many in the park. As the many glaciers melt, they form rivers that carry the water downward rapidly to the plains below.

194–195
Autumn in Jasper National Park brings the yellow of aspen leaves and larch needles. The gorgeous warm colors of fall will soon change enormously, becoming the cold whites, blues, and grays of winter. This high alpine wonder includes the very best Rocky Mountain scenery.

Each autumn the lower reaches of the parks begin to glow. Aspen leaves turn yellow, shimmering and vibrating in the wind. The larches, deciduous members of the conifer family, turn a brilliant golden hue before shedding their needles. The yellow trees ring evergreen layers below the high blue and white, snow-covered mountains. The autumn winds blow cold and animals prepare their burrows for the ferocity of winter. Gray skies turn the gorgeous high mountain lakes to bleakness, while the breezes kick up whitecaps on their rough surfaces. It will be many months before the tender buds come out on the trees, the valleys are once more carpeted with green, and the warmth of summer touches the meadows to allow wildflowers to color the landscape. This is the beautiful and harsh world of the Canadian Rockies.

With great foresight and wisdom, Canadian officials created vast parks and reserves in the Rocky Mountains of Alberta, preserving the ecosystem in the midst of so-called progress and ensuring its survival for generations to come. These parks are great tourist attractions, with Banff, the most visited, averaging four million visitors each year. The stress on the environment that results worries many conservationists, who want to ensure that the parks survive as unaffected by man as possible.

Banff is Canada's oldest National Park, created in 1885. Its 2,564 square miles of land encompasses twice the area of Yellowstone. The park was inspired by the completion of the Canadian Pacific Railroad through the region, and the railroad was the first means of bringing tourists to the site. Banff is accented by the quiet, placid Moraine Lake, the Valley of the Ten Peaks, and beautiful Lake Louise, with Victoria Glacier shining above and reflected in the lake's glassy surface. Today, Banff combines with Jasper, Kootenay, and Yoho national parks and three provincial parks to form the Canadian Rocky Mountain Parks World Heritage Site. A vast amalgamated preserve of eight thousand square miles, the region includes massive mountains of rock, glaciers, ice, snow, evergreen trees, clear running streams, and high mountain lakes.

The other major national park of the region is Jasper. Vast and mountainous, it is the largest and furthest north of Canada's National Parks in the Rockies. Jasper contains 4,200 square miles of incredible alpine scenery. Forty percent of the park lies above the timberline, a rugged country comprised of

195 Bottom right
Bighorn sheep make their way up a mountainside in Jasper National Park. The amazing bighorns nimbly jump from ledge to ledge, with specialized hooves that act like shock absorbers. Males stand 40 inches at the shoulder and weigh up to 350 pounds. The large curled horns grow larger on the male, up to fifty inches long, while the female's average fifteen. During the mating season, fierce battles take place between the males.

195 Bottom left
Roche Miette, a wonderfully eroded exposed granite monument, stands watch over the Rocky River, Jasper Lake, and a popular hot spring on the eastern side of Jasper National Park. The Miette Hot Spring averages a water temperature of 130 degrees Fahrenheit.

Lake Louise in Banff is one of the world's most scenic and beloved locations. The enormous blue-green lake reflects the Victoria Glacier and the high, tree-covered hills and rocky slopes behind.

JASPER AND BANFF NATIONAL PARKS

twisted and folded rock. Blue-white glaciers rise above thick evergreen forests and deep blue lakes. Jasper is far quieter and less visited than Banff.

Jasper is like two separate parks, divided roughly at the middle near the town of Jasper. The southern half is more heavily visited and has most of the park's attractions in a landscape composed of incredible peaks and glaciers. The northern half has fewer roads and trails, and the mountains are less spectacular, but more rustic and remote. Jasper was created in 1907 to stave off land speculation and development in the region.

Adjoining Jasper to the west in the Province of British Columbia is Mount Robson Provincial Park, which includes the

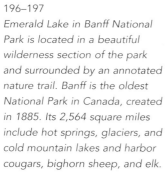

196–197

Emerald Lake in Banff National Park is located in a beautiful wilderness section of the park and surrounded by an annotated nature trail. Banff is the oldest National Park in Canada, created in 1885. Its 2,564 square miles include hot springs, glaciers, and cold mountain lakes and harbor cougars, bighorn sheep, and elk.

stark and cloud draped 12,931-foot Mount Robson, tallest of the Canadian Rockies. From the peaks called Whistlers in Jasper National Park one can get a marvelous view of the Athabasca and Miette river valleys to the east and to Mount Robson towering over all to the west. Another fascinating part of Jasper is Mount Edith Cavell, named for a brave nurse who died while serving with the British forces in the First World War. The landscape of Mount Edith Cavell lay beneath a glacier that disappeared less than one hundred years ago, and the effects of the glacier and the character of its alpine meadows, spectacular with wildflowers, can be studied today. Above, Angel Glacier flows out of a large bowl carved into a neighboring mountain. Below, clear, cold mountain lakes, sculpted by glacial activity, glisten in the sun. At the incredible Athabasca and Sunwaptha Falls, water pounds through narrow rock channels, while the Maligne River can be seen as it courses a limestone gorge toward the incomparable Maligne Lake.

197 Bottom left
Lake Peyto, seen here, and Lake Bow flank Banff's Bow Pass at 6,787 feet. At the lake one is surrounded by incredible scenery with plenty of ice, snow, and exposed rock. The pass may be covered with snow as late in the season as May.

197 Bottom right
The towering pyramid of Mount Athabaska, 11,870 feet tall, is the highest point in Banff. Autumn larch trees set off the white snow of the incredible massif.

miles, from central New Mexico to British Columbia. In the Canadian Rockies, the peaks are two to three thousand feet lower than those to the south in the Colorado Rockies, but the treeline is also proportionately lower, exposing more of the craggy alpine terrain. Since so much of the park's land is above the treeline, visitors are treated to views of Arctic

Altogether, the combined Canadian Rocky Mountain Parks have incredible scenery, with some of the best and most beautiful mountain views and alpine meadows in the world. It is a credit to the Canadian people that these natural wonders are being preserved in their incomparable majesty for generations to come.

198–199
The mountain peaks of Banff National Park, with their first dusting of snow in September,

beauty of this portion of the Rocky Mountains. The cool, clean air of the region combines with invigorating terrain to provide a

198 Bottom
The Athabasca Glacier in Jasper National Park spreads downward from the mountain to form the

Athabasca River. In the foreground, delicate tundra vegetation grows during the short summer season.

200–201
The spectacular Grand Canyon of the Yellowstone plunges one thousand feet from the plateau to the river below. The canyon has been rapidly eroded

during concentrated periods of time in the past, perhaps by huge floods caused by melting glaciers. Relatively little erosion takes place today.

200 Bottom left
Yellowstone protects many types of wildlife, including the majestic American elk, or wapiti. Each autumn, as male elks grow antlers and compete with one another for the favors of the females, the haunting sound of their bugling echoes throughout the park.

200 Bottom right
Bison graze in a field in Yellowstone, remnants of some forty million animals who once roamed America. Yellowstone National Park's bison herd, augmented by protection, was one of the small remnants of this animal species left alive in the early twentieth century.

GREENLAND

NORTH
AMERICA

WYOMING

PACIFIC
OCEAN

ATLANTIC
OCEAN

201 Bottom
Yellowstone is one of the few remaining habitats of the grizzly bear in the continental United States. A large, vicious, solitary animal averaging seven feet in height when standing on its hind paws, the grizzly is the picture of the wild American west.

NATURE'S WONDERLAND

YELLOWSTONE NATIONAL PARK

Mountains ring the horizon, while below, in the midst of a thick evergreen forest of lodgepole pines, a placid, meandering river suddenly turns into a roaring sheet of white water, flowing at an incredible pace down the turning, twisting canyon below. But more than even the waterfall, the character of the deep canyon with ochre, red, and orange walls catches the attention of viewers. This, the Grand Canyon of the Yellowstone, is but one of the myriad wonders preserved in this oldest of national parks.

Yellowstone National Park was established in 1872, and its formation began the modern, worldwide national park movement. It was one of the first times in history that land—in this case 3,474 square miles of it—was set aside by a government for the "benefit and enjoyment of all its people." The early explorers of this region agreed that it should never belong to just one person, or be owned or developed by anyone.

That attitude is easier to understand once one has seen Yellowstone. It is an incredible place with a many marvelous geological features packed into one region. An elliptical volcanic plateau that is ringed by mountains encompasses Yellowstone. The wonders of Yellowstone are the result of three volcanic eruptions, the latest of which took place 600,000 years ago. In fact, Yellowstone is a caldera measuring twenty-eight by forty-seven miles, one of the largest in the world. Beneath the forests and meadows, geothermal heat creates over ten thousand hot springs, fumaroles, mud pots, and geysers. Underground heat melts the winter snow, allowing the abundant wildlife of the park to sustain themselves on winter grasses. These animals include an estimated thirty thousand elk and twenty-seven hundred bison. The altitude of the area ranges from five thousand to over ten thousand feet at the tops of the mountains.

Yellowstone sits on the Continental Divide, which means that water from its northern half flows toward the Missouri River and the Atlantic, while water from the southern half flows to the Pacific. The Yellowstone River flows into Yellowstone Lake, the largest natural freshwater lake above seven thousand feet in the United States. It is twenty miles long by fourteen miles wide, with a shoreline perimeter of 110 miles, and was created by glacial activity, which sculpted the raw volcanic materials.

The Yellowstone River exits the lake and courses southward to tumble over Upper Yellowstone Falls. After a short respite, the Lower Yellowstone Falls send the water plunging 308 feet further down into Yellowstone's Grand Canyon. The spectacular canyon plunges one thousand feet from the plateau to the river below. The canyon's bright colors were created by hot underground water acting on the area's volcanic rock, causing a chemical reaction.

Nature puts on another important show in Yellowstone because of its enormous geothermal activity. Hot springs and geysers are created by surface water seeping down into porous rock where it is heated under pressure and rises back to the surface. Superheated water creates both geysers and hot springs. When the pressure drops suddenly the water

202 Right
Old Faithful Geyser is seen here
at sunrise. There are geysers and
hot springs in several geyser
fields at Yellowstone. Both
phenomena are created by
superheated water, except that
when the pressure drops suddenly
the flume of a geyser is created.

203
Visitors walk a narrow but defined
trail along the edge of the Grand
Prismatic Hot Spring in
Yellowstone. The spectacular

colors are created by
photosynthetic bacteria. A
combination of incredible
wildlife and unbelievable
geothermal phenomena makes
Yellowstone unique.

204–205
Yellowstone in winter presents an
entirely different experience to
the visitor. With fewer crowds,
areas such as the Firehole River
by the Lower Geyser Basin can
be enjoyed to their fullest
extent.

flume turns into a geyser's steam. Yellowstone has many
magnificent geysers, but the most famous is the one called
Old Faithful. This geyser earned renown for its regularity, and
currently erupts about every eight-one minutes with amazing
punctuality. Water is thrown 100 to 180 feet in the air. Other
geysers in the park are less regular but more powerful, some
throwing streams of white water as high as 250 feet.

Unlike geysers, hot springs occur when water that is not
superheated or under pressure emerges from the earth. This
water can also create fumaroles, which lack enough moisture
to flow and just vent steam; and mud pots, which form over
fumaroles as acid gasses decompose rocks into mud and clay.
The geothermal activity is kept going by a hot spot deep in
the earth's mantle. Magma is sent toward the surface in the

Yellowstone caldera, producing thirty times more heat
underground than is normal for this region.

Another incredible aspect of Yellowstone is its wildlife. The
park contains the largest concentration of mammal species in
the United States outside of Alaska. In addition to its famous
bison herds, Yellowstone also has huge herds of elk, nimble
bighorn sheep and mountain goats, and grizzly and black
bears. In addition to other forms of wildlife, the most popular
recent addition to the park has been the gray wolf.
Eradicated in Yellowstone in the early 1900s, the wolves have
been reintroduced to the park in a controversial program.

Yellowstone is not all sound and fury, not all spectacular
waterfalls and geothermal activity. Gorgeous open meadows,
winding clear streams, and beautiful evergreen-shrouded
mountains link these things. It is the complete ambiance of
Yellowstone, not just its spectacular elements, that makes the
park so special.

When wildfires raged across 1.4 million acres in and
around Yellowstone in 1988, people around the world were
made aware of this spectacular treasure of nature. Luckily, the
damage has been quick to heal, especially since fire is a
natural process in healthy forests like this one. However, the
blaze helped people realized once again what an incredibly
beautiful place Yellowstone is, and what a loss it would be if it
were ever marred in any way.

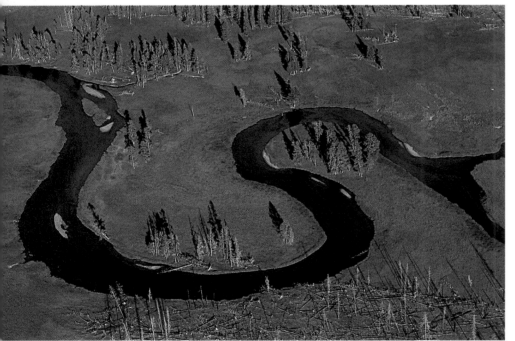

202 Top left
The Lower Yellowstone Falls send
water plunging 308 feet into the
Grand Canyon of the Yellowstone.
The river got its name not from this
canyon with its yellow rock walls, but
from the color of the rock near its
confluence with the Missouri River,
671 miles to the east in North Dakota.

202 Bottom left
The wondrous world of Yellowstone
National Park can be glimpsed in
this view taken along the winding
Yellowstone River. A world of
geysers, hot springs, meadows,
streams, lakes, and wildlife has
been preserved for the enjoyment
of visitors from around the world.

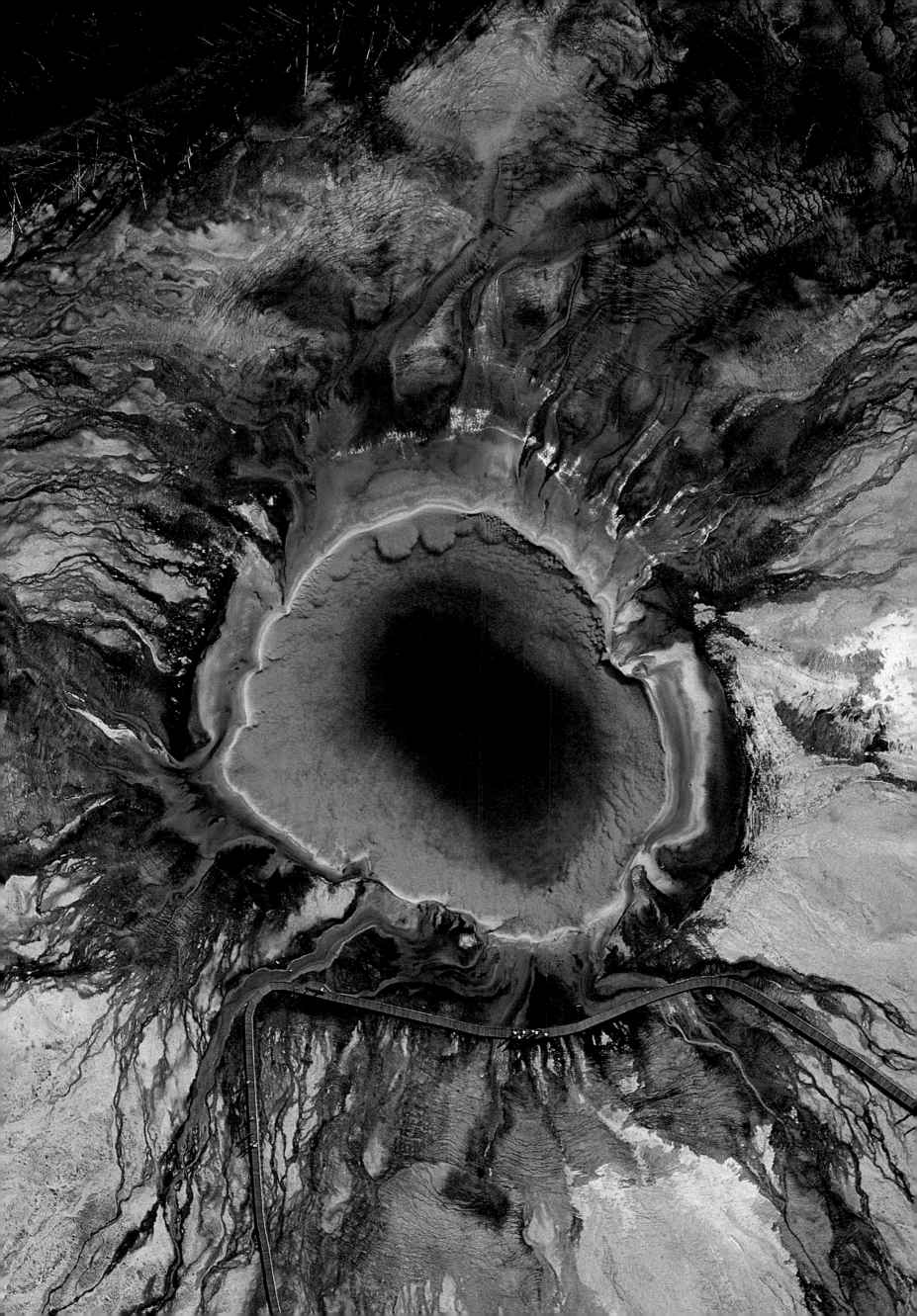

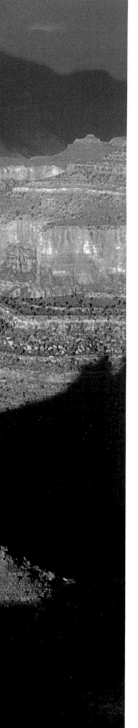

ARCTIC
OCEAN   GREENLAND

NORTH
AMERICA

PACIFIC
OCEAN   ARIZONA
                      UTAH
                NEW MEXICO
                                    ATLANTIC
                                     OCEAN

SILENT MAJESTY

THE CANYONLANDS OF THE AMERICAN SOUTHWEST

Ravens soar on silent wings high above the red rock walls. The huge opening in the earth below looks like something out of the prehistoric past, with rust-red layers extending downward in steep but graceful curves to a green and ochre plateau. Beyond the lip of the plateau, another drop leads to the Colorado River more than one vertical mile below. This is the Grand Canyon, one of several amazing natural features of the desert regions of the United States.

The Grand Canyon of the Colorado River in the state of Arizona is one of nature's masterpieces. President Theodore Roosevelt, who led the movement for the conservation of United States natural resources at the dawn of the twentieth century, advised Americans not to try to change it or mar it with man-made structures. By setting it aside in 1908, Roosevelt ensured that it would remain unmarred as a National Park, which it became in 1919. Today, Grand Canyon National Park encompasses 1,900 square miles. The North Rim is on the average three hundred feet higher than the South Rim , and is composed of spruce fir forest. Backpacking, hiking, river rafting, and enjoying nature's beauty are favored activities. And perhaps the area affords more—contemplation of the place humans occupy in a universe that includes something as massive and endlessly fascinating as the Grand Canyon. For in the end, the Grand Canyon is humbling. Like just a few other places on earth it illustrates how insignificant humans can seem in the midst of nature.

Visitors are inevitably awestruck when they first glimpse Arizona's Grand Canyon of the Colorado. They marvel at the huge twisted shapes of the mesas and mountains within the canyon itself, the breathtaking panorama of something so unimaginably large. They inevitably ask how the place could have been created. Scientists have been trying to answer that puzzle for many years. The Grand Canyon, probably the world's most spectacular example of erosion, is a chasm that follows the river for 277 miles and is up to eighteen miles wide in some places. Although the view from the rim is spectacular, nothing compares with walking down into the canyon to try to grasp its vastness and complexity. A hike

207 Center
Below Yavapai Point near Grand Canyon Village on the South rim, seen here, the canyon bottom is 2,400 feet above sea level and 4,500 feet below the rim—nearly a vertical mile. President Theodore Roosevelt declared the canyon a National Monument in 1908.

207 Bottom
The first human beings to live in the Grand Canyon were ancient Indian people. Seasonal hunters and gatherers, they spent summers on the rim and winters in the canyon. They left behind the caves they lived in as well as ruins of structures including granaries.

208 Top left
A towering rock formation in Monument Valley is a testament to the erosive forces of nature. Many Navajo (Diné) people live in the valley, some in hogans, the traditional mud and log homes of their people.

The Navajo who live in Monument Valley are among the most traditional of their people. Few have running water, electricity, or any modern conveniences. They herd sheep and goats; some weave rugs and make beautiful silver objects.

208 Top right
About two hundred million years ago, in the late Triassic Period, the region of Petrified Forest in Arizona, seen here, was a vast floodplain crossed by streams.

208 Bottom
Delicate Arch is one of over 1,500 catalogued arches within Arches National Park in Utah. This incredible area of the U.S. desert contains geological marvels created by wind, water, extreme temperatures, and underground salt movement. Delicate Arch is an isolated example of the erosion process, standing before the gorgeous La Sal Mountains.

208-209
Wind and water carved the incredible red sandstone mittens of Monument Valley.

209 Bottom left
Sand dunes combine with rock formations to create the memorable landscape of Monument Valley. The region has often been used as a backdrop for movies about the American West.

THE CANYONLANDS OF THE AMERICAN SOUTHWEST

takes one down through the layers of ancient rock, just as it proceeds down through six of the seven climatic belts. Vegetation varies from that of the Mexican desert on the canyon floor to that of alpine regions on the north rim.

It has been estimated that it took three to six million years for the canyon to evolve to its present appearance. Since most of the rock that comprises the canyon was formed as sediment beneath oceans, it is thought that the area of the Grand Canyon was once buried under a vast inland sea. Millions of years of the movement of continents, tectonic plates, and the advance and retreat of the oceans created the rock layers. The action of rainwater falling on the mountains caused streams to unite into a riverbed. This formed the ancient Colorado River, which began to carve its way through a flat plain. At the same time, the Colorado Plateau beneath it was actually rising due to pressures deep in the earth. This uplift can be compared to one person holding a board that another is sawing. It allowed the river to cut through the rock of the plateau very efficiently.

The result of nature's handiwork can be seen all along the length of the Grand Canyon. What water did not accomplish, wind, ice, ancient volcanism, and rockslides finished. The Grand Canyon is an amalgam of nearly every type of force that can be exerted upon rock by nature.

Similar forces created other wondrous places in the southwest United States. Lake Powell fills the naturally sculpted Glen Canyon, formed by the waters of the Colorado River above the Grand Canyon. Today, the canyon is dammed and buried beneath the blue waters of Lake Powell, which winds for 186 miles through the deserts of Utah and Arizona. There are literally hundreds of side canyons, inlets, and coves that harbor Indian ruins and fantastically shaped rock formations.

The hand of man tamed nature when Glen Canyon Dam was completed in 1966, putting a large body of water in the midst of the desert. The lake was named for explorer John Wesley Powell, who led the first team of men to successfully

209 Bottom right
Colorado National Monument in western Colorado is a semiarid region with interesting rock formations caused by erosion. Deep canyons etch the park, including the headwaters of the Colorado River.

210–211
Snow blankets a portion of the Great Basin in Nevada. Covering a large portion of Utah and Nevada, the Great Basin is a desert region created by the rain shadow of the towering Sierra Mountains to the west. Although it is a hardscrabble and desolate looking area, this flat desert, bounded by beautiful mountain ranges, has a charm all its own.

212 Top left
Sometimes the sweep of desert rock obscures the thriving plant life that makes its home near Lake Powell. For instance, lichens grow on the canyon walls, showing up as streaks of color. Cactus and bayonet-tipped yucca grow widely. Wildflowers bloom in the spring and even after a particularly heavy summer rainy season.

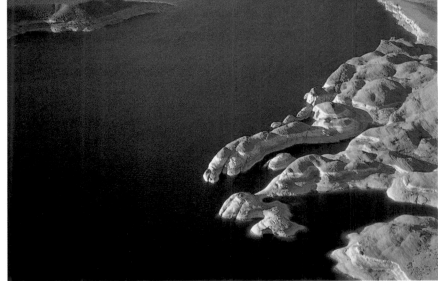

212 Bottom left
Although it may look barren, the desert near Lake Powell supports wildlife. Lizards like the chuckwalla are out in the daytime, as are snakes, for they depend upon the sun's heat to regulate their body temperatures.

212 Top right
The desert surrounding Lake Powell is home to many animals. Most do their hunting at night. Coyotes and foxes, ringtails, deer, rats and mice all conserve energy by holing up during the day. Many species of birds live in and about the canyon, including waterfowl drawn there since the advent of the lake.

212 Center left
Towering rock formations and soaring red cliffs compose the desert lands surrounding Lake Powell. The rugged area landscape is impressive and memorable, perhaps more so than the man-made lake in its midst. Away from the lake one can find quiet and solitude.

212–213
Lake Powell, formed by the waters of the Colorado River backed up behind the Glen Canyon Dam, winds for 186 miles through the deserts of Utah and Arizona. There are literally hundreds of side canyons, inlets, and coves that harbor Indian ruins and fantastic rock formations. The Glen Canyon Dam was built between 1956 and 1964 using five million cubic yards of concrete. Many environmentalists criticize the inundation of spectacular Glen Canyon.

navigate the length of the Colorado River canyons. Ironically, the reservoir has completely changed the landscape and the beautiful Glen Canyon from what Powell saw in 1869.

Towering rock formations and soaring red cliffs compose the desert lands surrounding Lake Powell. The rugged landscape is impressive and memorable, much more so than the man-made lake in its midst. Away from the lake one can find quiet and solitude. The graceful red arch of Rainbow Bridge, which at 290 feet tall is the world's largest natural bridge, is probably the most spectacular formation near the lake. Rainbow Bridge is a sacred place for the Navajo (Diné) Indian people, whose reservation borders the south side of the lake.

North of the Grand Canyon in Utah is Bryce Canyon, which covers 37,277 acres of scenic, colorful, and unique rock formations called hoodoos. The rock is limestone, sandstone, and mudstone, carved from the eastern edge of the Paunsaugunt Plateau in southern Utah. The hoodoos were created ten million years ago by uplift and faulting, when one set of stone blocks was thrust upward and another became the Paunsaugunt Plateau. Ancient rivers sculpted the tops of the upper blocks, and with time tall thin ridges called fins emerged. Sculpted by wind and ice, the fins eroded into the amazing hoodoos one sees today. The red hoodoos create spectacular views when seen with the plateau's pine forests behind them. In the winter, snowfall makes the hoodoos look like confections and presents a rainbow of natural colors. Bryce Canyon is a wonderful spot to hike and contemplate nature's wonders.

The desert is home to many animals, most of whom do

213

Like the towers of castles, Lake Powell's blue waters surround the red sandstone mesas of the southwestern desert. Lake Powell is a National Recreation Area administered by the U.S. National Park Service. The most popular

form of recreation on the lake is probably boating, and on any given day powerboats, canoes, kayaks, sailboats, and even houseboats can be seen plying the waters.

their hunting and moving about at night. Coyotes and foxes, ringtails, deer, rats, and mice all conserve energy by holing up during the day. Lizards like the chuckwalla are out in the daytime, as are snakes, for they depend upon the sun's heat to regulate their body temperatures. Many species of birds live in and about the canyon, including golden eagles and ravens, which soar on the winds above.

Sometimes the sweep of desert rock obscures the thriving plant life that makes its home in the desert. For instance, lichens grow on the canyon walls, showing up as streaks of color. Cactus and bayonet-tipped yucca grow

YOSEMITE NATIONAL PARK

El Capitan and its sheer walls provide an almost irresistible challenge to rock climbers. And it is a rock—a solid boulder of granite rising 3,604 feet from the floor of the Yosemite Valley. El Capitan is the largest single outcrop of granite in the world. Halfdome, one of the most photographed landmarks in the world, is located across and down the valley from El Capitan. Rising 8,852 feet, Halfdome, beloved of John Muir, Ansel Adams, Theodore Roosevelt, and millions of other less famous visitors, protrudes through the cloud cover over the Yosemite Valley.

Yosemite is also a land of water. Spectacular waterfalls seem to be everywhere in the valley. Two of the world's ten tallest waterfalls are located in the Yosemite Valley, Yosemite Falls and Ribbon Falls. With a combined drop of 2,425 feet, Yosemite is the third tallest waterfall in the world. Nevada Falls is located on the Merced River, which plunges over the valley wall, collects at Emerald Pool below, then tumbles down 317 feet in Vernal Falls toward the valley floor. All of the waterfalls are spectacular long ribbons of white water, foaming and crashing to pools below the heights.

In addition to the treasures of the Yosemite Valley, the park harbors an amazing ecosystem with some 1,300 types of flowering plants, 31 species of trees, 220 species of birds and 60 mammal species, of which deer and black bear are the most common. One of the most unusual and special of the park's species is its giant sequoia trees. The sequoias grow in a narrow band that stretches along 260 miles of the Sierras and is limited by elevation; the trees can only grow between 4,500 and 7,500 feet. Today there are about thirteen thousand of these trees, growing in seventy-five groves throughout the mountains. The trees live for several hundred years and are among the world's oldest living things.

222 Top
One of the richest underwater marine habitats in California lies off the Point Lobos State Reserve along the California Coast. In the ocean waters, amid sea kelp forests that grow to a height of seventy feet, are animals such as seals, sea otter, and whales.

222 Bottom
Waves crash onto the shore below the Pacific Coast Highway in the Big Sur region in this view, which shows the interior's deeply incised canyons. Point after point of rocky land can be seen as one looks up and down the coast.

222–223
One of the spectacular, graceful bridges that pass over the deep canyons along the California coast can be seen in this photo of the Big Sur region. South of Carmel is a beautiful stretch of Route 1, the Pacific Coast

Highway, which has become popularly known as the Big Sur Country. There are few towns along the highway, only scenic views and twisting, winding roads that follow the contours of the coast and its canyons.

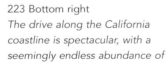

223 Bottom right
The drive along the California coastline is spectacular, with a seemingly endless abundance of

new vistas with every turn. One of the few towns along the route is Big Sur, from which the name of the region was derived.

223 Bottom left
The gorgeous California Coast runs for much of its length in an undeveloped state. The area is rich in natural beauty and has an

abundance of animal and marine life. Sea lions, harbor seals, elephant seals, sea otters, and whales and dolphins are seen in its waters near the reserve.

224
The Pacific Ocean pounds the shore of California's Santa Cruz Island, while sea birds wheel and call overhead. Santa Cruz is about twenty-four miles long and comprises ninety-six square miles. Several areas on the island provide important habitat for nesting sea birds, plants, and animals. More than six hundred plant species and 140 land bird species live there. The cliffs along the shoreline, offshore rocks, and tidepools all provide breeding habitats.

CALIFORNIA'S EXTRAORDINARY COAST

225 Top left
The Channel Islands comprise an island chain lying off the south coast of California in the Pacific Ocean. One of eight islands, Anacapa Island lies eleven miles off the coast from Oxnard. It is actually composed of three islands, seen in this view, about five miles in length and comprising 699 acres. Anacapa Island has no fresh water.

225 Top right
Anacapa Island is a place of great scenic beauty. To the uninitiated it looks barren, yet it supports many species of plants. Biologists enjoy studying the ecosystem of the Channel Islands, where many plant and animal species have adapted to the special conditions and differ from similar species on the mainland.

225 Center
The romantic cliffs near La Jolla along the Pacific Coast are located just north of San Diego. The rugged Pacific Coast draws many visitors who watch seabirds and the ocean waves. La Jolla means the Jewel in Spanish, and the community offers beaches, cultural activities, and fine restaurants.

225 Bottom
The eroded cliffs along the coast near La Jolla epitomize the romantic charm of the California coast. Areas of awesome beauty are joined with a southern European climate to make the region one of the most beautiful in the world.

California coast are located on the Channel Islands. Five of the eight islands and the surrounding nautical mile of ocean are parts of Channel Islands National Park. Santa Cruz Island, which is twenty-four miles long and 60,645 acres, is the largest in the group. The highest of the Channel Island's mountains is found on Santa Cruz, standing at 2,470 feet. Santa Cruz is a place of great scenic beauty, with two rugged mountain ranges, deep canyons, a wide central valley, sea cliffs extending for seventy-seven miles, huge sea caverns, tidepools, and beaches. The cliffs along the island shorelines, offshore rocks and tidepools provide breeding habitat for more than 600 plant and 140 land bird species. Their rough, seemingly barren topography supports unique plant and animal species that have adapted to special conditions that differ from those that similar species on the mainland require. Channel Islands National Park is part of the International Man and the Biosphere program, and it is also the center of a National Marine Sanctuary.

Below Los Angeles along the California coast one enters a climate reminiscent of the Mediterranean. The rugged Pacific coast off La Jolla draws many visitors who enjoy the area's sandy beaches at the head of canyons and watching sea birds fly above as ocean waves crash against rocks. Each winter the telltale waterspouts of gray whales can be observed offshore beyond the kelp beds, in the midst of an annual five-thousand-mile migration.

At San Diego, Cabrillo National Monument is located on the Pacific side of the San Diego Bay and has tidepools where all manner of sea creatures can be observed at low tide. Favorites include the flowery anemone, the lined shore crab, and limpets. The tidepools lie below rough craggy black rocks that are submerged during high tide. The slick black rocks in the tidepools are particularly inviting at sunset, when the sinuous lines of the cliffs above have an orange glow. In addition to the gorgeous views from the cliffs at Cabrillo, the white sandy beaches of the long Coronado Peninsula to the south make San Diego a special place, in love with and a part of the sea itself.

The entire length of the California coastline is warm and romantic. Pounding waters, gorgeous sunsets, dramatic cliffs, fascinating wildlife, and endless opportunities for water sports make this region one of the world's most inviting natural wonders.

230–231
A cloud of spray rises from the thundering torrent of Niagara Falls, located along the border of the United States and Canada between Lakes Ontario and Erie

on the Niagara River. In this view, the boat Maid of the Mist approaches the Canadian, or Horseshoe, Falls, to give tourists a good look at their extraordinary power.

230 Bottom left
This aerial view of the Canadian Falls shows the expanse of the Niagara River approaching the precipice, with Goat Island on the left. The water flows out of the

Great Lakes, and through Lake Erie to get to this point. A few miles north of the falls, the Niagara River empties into Lake Ontario and from there flows down the St. Lawrence River to the Atlantic Ocean.

231 Bottom
The raging force of Niagara Falls flows over the escarpment at the rate of 194,940 cubic feet per second. The Canadian Falls are

eroding at the rate of five feet per year. The American Falls, which carry a smaller volume of water, are eroding at only six inches per year.

230 Bottom right
The reason for the name "horseshoe" is evident in this dramatic aerial view of the Canadian side of Niagara Falls. The horseshoe shape is 2,592 feet

long and drops 160 feet to the river below. The American side, seen at the right of the photo, is on the western border of New York State and slightly higher but is only 1,001 feet along the escarpment.

GREENLAND

NORTH
AMERICA

ONTARIO
NEW YORK

PACIFIC OCEAN

ATLANTIC OCEAN

AN IMMENSE
BOILING CAULDRON

N I A G A R A   F A L L S

You can walk right up to the edge, grasp an iron rail, and watch as a raging torrent of water flows over the escarpment at the rate of 195,000 cubic feet per second. It flows with such volume, with such force, that you feel that you are a part of it as you stand near the edge. The rapids above turn into an irresistible flow of green, blue, and white water falling to the swirling currents below. There are few places on earth where such palpable power can be felt.

And people have been feeling that power for hundreds of years. The American Indians of the region knew of this mighty scene and showed it first to Jesuit missionaries. By the nineteenth century, tourists from around the world made Niagara a must-see American destination. In 1844, a British traveler who saw Niagara Falls for the first time spoke of the "cloud-like smoke which overhangs the Falls, and gives to the white foam below the appearance of an immense boiling cauldron, and when the sun shines on the falling spray the ever-changing prismatic colors, so well described by the Poet, may be observed." Niagara Falls soon became the prime tourist and honeymooning destination. Towns grew up on both sides of the falls, early photographers tried to capture their majesty on slow Daguerreotypes, and wirewalkers and men in barrels tried to survive and get their names into the history books.

In short, Niagara Falls is one of the most accessible of nature's wonders. They occur almost midway along the Niagara River, which flows from Lake Erie into Lake Ontario, draining the huge Great Lakes of North America. The falls formed about twelve thousand years ago, when glaciers retreated northward and water from Lake Erie began flowing over the Niagara Escarpment, a ridge that runs from New York to Wisconsin through the Great Lakes region. Erosion has since pushed the falls eleven miles upstream, forming the spectacular, steep and rocky Niagara River Gorge below the falls. Today, two nations share the falls, Canada on the west and the United States on the east. An "International Peace Bridge" connects the two nations, which meet here as equals.

But the falls are not equal in their beauty or the views one receives of them. For Niagara Falls is really two separate waterfalls side by side, separated by Goat Island in the middle of the Niagara River. The Canadian side of Niagara forms a graceful horseshoe shape 2,592 feet long, and drops 160 feet to the river below. The American side is slightly higher, but measures only a straight 1,001 feet along the escarpment. The Canadian falls are eroding at the rate of five feet per year. The American Falls, which carry a smaller volume of water, are eroding at only 6 inches per year.

A couple of miles downstream from the falls, an enormous whirlpool spins in a blue-black arc, while man-made power plants generate electricity from the force of the river's wash.

The thundery water pouring over Niagara Falls may be viewed from several different vantage points. The adjoining towns of Niagara Falls Canada and its counterpart in New York provide tourist facilities, hotels and museums about the magnificent waterfall. Parks on each side of the falls allow visitors to stand close to the thundering edge, observation towers allow a view from above, tunnels lead down to observation platforms at the base of the falls, while boats called "The Maid of the Mist" take visitors as close to the crashing base as they can with safety. At night, the immense waterfall is illuminated by huge xenon lights of various colors, creating a gorgeous if artificial display of Niagara's beauty.

But commercialism and a determined tourist trade cannot ruin either the power or the beauty of the falls. A visitor today can stand in the same spot as a British visitor in 1844 and see the same awesome, breathtaking view of this timeless wonder.

CENTRAL AND SOUTH AMERICA

INTRODUCTION

232 Top
The incredible mist-shrouded
falls called Iguacu, or Great
Water, are located near the
junction of three nations,
Argentina, Brazil, and Paraguay.

Iguacu is the second widest
waterfall in the world after
Victoria Falls in Africa. Their
incomparable beauty
complements the lush forest
that surrounds them.

The beauties of South America range from the beaches of Brazil to the tops of the Andes, from watching penguins at the tip of Argentina to watching exotic birds on a Caribbean island. The tropical rainforests of the Amazon basin have still not been fully explored, with flower and insect species still unknown.

South America is the fourth largest continent, and a large portion of its landmass is still relatively wild and unpopulated. In fact, the recent movement of population into the interior of the rainforests, burning and cutting down the thick vegetation, has become a major cause of concern.

South America has some of the world's largest deposits of minerals, timberlands, and farmlands. It also has incomparable beauty, which has become a marketable commodity in recent decades in areas that formerly had no appeal. The advent of eco-tourism—viewing areas with abundant or rare bird, plant and animal species—has become quite popular. The Amazon rainforest, the high Andes, and the southern latitudes of Argentina and Chile have become increasingly popular eco-tourism destinations.

In addition to its natural wonders, South America also has a significant cultural history. The continent was home to the most advanced Native American culture, the Incas of Peru. After the Spanish and Portuguese became aware of the continent in the early 1500s, they colonized it, exploiting it for its natural resources. Inspired by the American Revolution, most of South America shook off the yoke of colonialism during the twenty-year period from 1810 to 1830, setting up the independent republics of Paraguay, Argentina, Chile, Columbia, Peru, Brazil, Ecuador, Bolivia, Uruguay, and Venezuela.

More than three quarters of South America lies in the tropics, as the equator crosses the continent at its widest point. Water surrounds it, except for the narrow Isthmus of Panama on its northwest corner. South America has four major land regions. The Pacific coastlands form a very narrow strip

232 Bottom
The jaguar is one of the most
beautiful forest animals of South
America and one of the most
feared, though it rarely attacks
humans. Jaguars feed on all
types of animals and are

considered nuisances by South
American cattle ranchers. It is
the largest of the American
members of the cat family, with
adults measuring six feet long
with an additional thirty inches
for the tail.

232–233
The broad, low-lying plain of the
Amazon River valley floods from
November to May each year,
when streams of fresh water from
the Andes Mountains to the west
pour billions of gallons of water

into the wide river. Heavy rains
also contribute to the flooding.
The fertile rainforest is inundated
in sections, but plants and
animals take it all in stride. Water
is the lifeblood of the Amazon
rainforest.

between the Andes on the east and the ocean on the west.
They range from five to fifty miles in width and vary from
swampland in the north in Colombia to desert lands in Peru
and Northern Chile, to fertile farmland in central Chile, to
stormy fjords along the last thousand miles of the southern
coast. The Andes Mountains form the backbone of the
continent and run along the entire Pacific Coast from north to
south, their tall peaks adding a wonderful white coda to the
horizon. The central plains to the east of the Andes constitute
about three fifths of the continent. This region includes
grassy plains called llanos in Colombia and Venezuela,
tropical rainforests in the Amazon River basin, forested
scrublands in the Gran Chaco region of Paraguay and
Argentina, and ranch and farm land called the pampas in
Argentina. Geologically, the Eastern Highlands are similar to
the North America's low Appalachian Mountains. They run
from the Guiana Highlands to the Patagonian Plateau in the

south. The climate ranges from tropical in the Amazon basin
to the frigid weather of the high Andes. But most of the
continent is warm or hot year round.

Because so much of South America is relatively unspoiled,
it is the home of magnificent natural areas of great interest.
The tropical rainforest of the Amazon basin includes many of
these wonders, including incredible waterfalls and
concentrations of animal and plant life. The study of animals
and their evolution and adaptations is a major interest in the
Galapagos Islands, which are located six hundred miles off
South America, and on Antarctica, the continent to the south
of Tierra Del Fuego. The beautiful islands of the Caribbean
Sea have become important as resorts for urban dwellers of
several continents.

Many South American countries have been diligent in
preserving their natural wonders, and National Parks have
been set up in various regions of South America to ensure

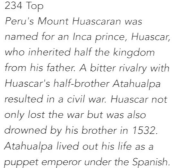

234 Top
Peru's Mount Huascaran was named for an Inca prince, Huascar, who inherited half the kingdom from his father. A bitter rivalry with Huascar's half-brother Atahualpa resulted in a civil war. Huascar not only lost the war but was also drowned by his brother in 1532. Atahualpa lived out his life as a puppet emperor under the Spanish.

234 Center
One of the most astounding sights in the world, Machu Picchu, an ancient city of the Incas, stands in ruins atop a mountain ridge in Peru. Built about 1200 A.D., the method of construction still baffles and amazes historians and engineers.

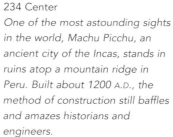

INTRODUCTION

the preservation of these landmarks. In addition to the rainforest, the other major natural wonder of South America is the Andes mountain chain, the longest in the world.

The wildlife of the continent is varied and often exotic, with thousands of bird species, including colorful tropical birds, many varieties of monkeys, and animals that will forever be identified with South America—llamas, alpacas, vicunas, chinchillas, tapirs, capybaras, anacondas, marmosets, jaguars, piranhas, condors, and sloths. Almost a quarter of all the known animal species live in South America, not to mention the untold numbers of insect species that thrive there. South America's plant life is equally rich and varied. As a result of the tropical climate and long growing season, plants thrive throughout most of the continent and there may still be a vast number of species of plants that remain undiscovered.

These and other incredible resources make this a continent of vast importance to worldwide ecosystems, the global climate, and the world's biological diversity. One argument for preserving the rainforests is that they may contain a heretofore-unknown species or application of a plant or animal that may contain a cure for a presently incurable disease. This is certainly a valid reason to preserve something.

But it is more difficult to argue that the pure beauty and wonder of an area is reason enough to save it. With the growing eco-tourism business, perhaps South Americans will find that the best way to exploit their incredible natural resources is to show them off rather than destroy them through development. Many South American nations have already made a good start with preserving their natural wonders with their National Parks and World Heritage Sites.

234 Bottom
The spiked pinnacles of Argentina's Cerro Torre are seen here from the summit of 11,073 foot Fitz Roy, on the border of Argentina and Chile. Their dramatic coloring and the way in which the snow glazes the brown rock lures nature-lovers and photographers to the region. Located in the far southern part of both nations, less than five hundred miles from the tip of the continent, the mountains lie within the boundaries of Los Glaciares National Park.

234–235
The snowy peak of Mount Huascaran dominates the Peruvian Andes at 22,205 feet. A dormant volcano, the mountain is covered by the Cordillera Blanca Glacier. It is located just fifty miles inland from the Pacific Coast, and only two hundred air miles north of Lima. The surrounding country is extremely mountainous and few roads penetrate the region.

235 Bottom
High in the Cordillera Vilcanota of the Peruvian Andes, a hot spring near the Ausangate Mount attests to the hot magma bodies intruding and solidifying in the earth's crust. Earthquakes are evidence of the continued pressure and friction caused by tectonic plates.

238 Top left
Basse Terre, Guadelupe's minor island, is composed of volcanic mountains. It is divided from *Grand Terre, the major island, by a narrow strait called La Rivière Salée.*

hammerhead sharks, and flying fish. Caribbean waters generally have a visibility of up to one hundred feet, making these undersea wonders easy to enjoy.

The region's climate of tropical heat tempered by daily rains and cooling breezes creates an environment in which such a profusion of flora and fauna can live. Bright white sandy beaches contrast with the dark green foliage of the forests and the bright turquoise waters offshore to create an irresistible environment, one that is beautiful and inviting. Coves and hills covered with lush tropical vegetation ring with the calls of exotic birds and animals.

The wide, open bays of the islands, with their deep blue

238 Top right
Once the realm of pirates in tall ships who flew the "Jolly Roger," the British Virgin Islands are today a tourists' paradise. Located east of Puerto Rico, just sixteen of the thirty-six islands are inhabited and there is only one town of any size in the entire group, Road Town on the southeastern coast of Tortola.

238 Bottom
The fertile volcanic soil of Basse Terre, Guadalupe and the ample rainfall it receives make the island a botanical wonder. Topography ranges from tropical rainforest in the mountains to savannah along the sea. The island is home to a variety of birds, butterflies, lizards and iguanas.

waters, complement the tall, sharply-peaked and rolling green hills of the interior. Dark black lava rock outcroppings pockmarked with the fury of a lava flow accumulating over millions of years punctuate the white sands of the beaches. Colorful birds with long bills of red, green, yellow, and orange inhabit the trees above, while colorful and fragrant flowers and plants lend a cast of red to the underbrush. Hummingbirds flit from flower to flower, looking themselves like rare, vibrant jewels. At the end of the day, high on a volcanic outcrop, a gorgeous fiery red sunset turns the forest, the hills, the beach, and the ocean to gold. Darkness falls, but the beauties of the islands will be there again in the morning, while stars watch over them throughout the night.

The islands range from large masses like Cuba, Puerto Rico, Jamaica, and Haiti to small gems like Aruba, the Bahamas, Martinique, Guadeloupe, Barbuda, and the Grenadines. Each of these islands has its own charm, its own special characteristics that set it apart from the others. Jamaica's tall peaks, the Blue Mountains, rise more than seven thousand feet and provide lush terrain for cocoa and coffee plantations. The plateaus drop to forested valleys with mountain streams and waterfalls, spilling through green pastures and connecting to form rivers that tumble swiftly to the sea. A smaller island like Antigua includes a limestone and coral coastline scalloped with bays, jagged cliffs, sandy coves, and inlets. Its interior is composed primarily of dry and open farmland. Antigua's highest mountain is Boggy Peak, 1,319 feet tall. A large island like Puerto Rico includes a variety of ecosystems, including the long, rugged spine formed by the central mountain range, a 28,000-acre rainforest near San Juan, over three hundred miles of sandy coastline, a desert, and one of the world's largest river cave systems. The northern portion of the island is lush and green, while the south tends to be arid. The tallest peak, Cerro de Punta, rises 4,398 feet above the surrounding plain. All this and the world's third largest underground river, the Camuy, are found on one island.

It is this type of physical beauty and variety that make the Caribbean Islands, large or small, attractive to tourists as well as island residents. The people of the islands are beginning to realize the potential—and the pitfalls—of heavy tourism. Little could Columbus have imagined the bustling tourist towns of today, the crowded beaches and the rushing traffic along island roads. Some of the islands, like Guadeloupe, the Caymans, and Aruba, have set aside national parks to preserve their tropical forests and coral reefs, insuring that the wonders of today will be perpetuated for the future. Many also have established bird sanctuaries and wildlife refuges. Parts of the U.S. Virgin Islands and Puerto Rico are managed by the U.S. National Park Service, which preserves both historical and natural features. This dedication to the preservation of the region's fragile resources is fitting, for the Caribbean Islands are truly one of the world's great natural wonders, places well worth preserving for future generations.

238–239
Bottom Bay in Barbados is one of the few anchorages on this Caribbean island surrounded by coral reefs. Barbados is flat along the coast and rises to a hilly interior, its highest point being just 1,104 feet above sea level. An odd assortment of animal life inhabits the island.

239
The very end of the long peninsula of Pointe des Chateaux on the east side of Guadelupe's Grande Terre Island showcases its mostly flat topography. A French possession in the Lesser Antilles, Grande Terre has a dry climate and features sugar cane plantations in its limestone soil.

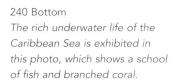

240 Bottom
The rich underwater life of the Caribbean Sea is exhibited in this photo, which shows a school of fish and branched coral.

Visibility is excellent, often exceeding one hundred feet, allowing the viewer to experience these wonders easily and with a sense of perspective.

240 Top left
A trumpet fish dives amidst hard branched coral beneath the waves of the Caribbean Sea. The unique, narrow fringing reefs of the Caribbean support a population of tropical fish, including the queen triggerfish, angelfish, butterfly fish, parrot fish, rays, and sharks.

240 Top right
A purple and yellow tube sponge is a fine example of one of the many forms of sponges, the most primitive of multicellular animals. Fossil sponges date back at least six hundred million years, to the Precambrian Period. The interior crevasses of these sponges provide a refuge for marine invertebrates like crabs and sea stars.

241
An undersea garden spreads before the eye in this view, which includes gorgonia sea fans, sponges, sea flumes, and hard branched corals. The perfect conditions of water temperature, salinity, clearness, and distance from the surface must be met before these organisms can thrive as they do here.

A queen angelfish is seen in this
view. Differing from butterfly fish
in that they have a spine near the
lower edge of the gill cover, most
angel fish are small but some
grow up to twenty-four inches.
Another popular variety, the
French angelfish, is identified by
its predominantly black color and
the prominent yellow stripes
down the sides of its body.

243 Top
A stingray floats, batlike, close to
the sandbed of the Caribbean.
Stingrays have small mouths with
blunt teeth that they use to crush
their food, primarily crustaceans.
The long whiplike tail, used
primarily as a rudder, also has
barbed spines that can inject
poison into a victim.

243 Center
A French angelfish displays the large dorsal and anal fins common to the species.

243 Bottom
A common and relatively harmless shark, a nurse shark, dives beneath a rock outcropping looking for prey. Sharks date back 100 million years, and are some of the most successful creatures on earth, true "eating machines" that have evolved to the point of startling efficiency in capturing their prey.

THE CARIBBEAN

CARIBBEAN SEA

VENEZUELA

SOUTH
AMERICA

ATLANTIC
OCEAN

PACIFIC
OCEAN

248 Top
*Canaima National Park, a United
Nations World Heritage Site,
preserves mesalike geological
formations called tepuis. Unique
geologically and biologically,
they are also extremely beautiful,
particularly when glimpsed with
a foreground of vegetation and
placid water, as in this scene.
The area is still rugged and
difficult to access.*

THE LOST WORLD

ANGEL FALLS, VENEZUELA

The aviator swooped his airplane low over the Venezuela jungle plateau, searching for the exact tabletop mountain where he had found gold months earlier. He had been in the region once before, but could not find the particular mountain he was looking for. An old prospector had hired Jimmy Angel, and the two had visited the top of the mountain, striking it rich but unable to carry off the gold in their light plane. The old man died before they could return, and now Angel searched, trying to find the exact spot once more. Finally, spotting a familiar landmark, he landed his plane, but his wheels sank into the mud of the mountain's flat top. In seconds, the man's plane became useless. It was 1935. U.S. aviator and would-be millionaire Jimmy Angel was stranded on top of a mountain hundreds of miles from any settlement, in the middle of what Arthur Conan Doyle described as the Lost World. It took Angel eleven days to walk out of the jungle; his airplane wasn't recovered until 1970. When Angel got back to civilization, just happy to be alive, it was scant consolation that he could report the existence of a waterfall previously unknown to the outside world. It was a ribbon of water that dropped from the top of the cliff on which his plane was stranded. Jimmy Angel didn't know it, but he had discovered the world's highest waterfall.

Located on Mount Auyan in southeastern Venezuela, Angel Falls has an uninterrupted drop of 2,648 feet, and a height of 3,212 feet. It is over a thousand feet higher than any other known waterfall. An expedition from the United States explored it in 1949, taking photographs and measuring it.

Auyan-tepui (Devil's Mountain in the language of the local Pemon people) is a table mountain, or tepui, and is shaped like the claw of a giant crab when seen from above. Angel Falls is located at the crux of the claw, down a long canyon. The area has been compared to the Lost World that Sir Arthur Conan Doyle described in fiction in 1912. Doyle based his novel on the reports of British botanist Everard Im Thurn, who explored the region in 1884. Doyle mused upon the idea that mesas existed in South America so remote and cut off from the rest of nature that evolution had been arrested or continued independently for millions of years. Perhaps dinosaurs or their evolutionary descendants might still live in such a place. Today, Doyle's story sounds far-fetched, yet in some respects southeastern Venezuela is, in many ways, a Lost World, for it supports a variety of unique plant and animal species atop its tepuis.

Angel Falls is located within Canaima National Park, a United Nations World Heritage Site. The park is spread over three million hectares in southeastern Venezuela along the border with Guyana and Brazil. The park is composed of roughly 65 percent tepui formations. The tepuis are gigantic, flat-topped sandstone mesas, and they are unusual geologically and biologically. Their sheer cliffs and waterfalls form a spectacular landscape. Fewer than half of the hundred known tepuis have been fully explored. The tepuis were created when fissures in the continental plate were eroded and then uplifted. The tepuis achieved their present appearance about four million years ago.

Scientists eagerly visit this region, unique because of its isolation and the incredible abundance of water. Some of the tepuis are actually tall enough to snag clouds and create their own weather, and rainfall averages about 150 inches a year. Perhaps as many as half of the species in the region are unique to the tepuis, a result of the incredible wetness of the area and their isolation on the tops of these mesas.

Angel Falls is located on the Caroni River, which flows into the Orinoco. The fall is most spectacular during the rainy season, from June to November. The majority of the water vaporizes and forms exotic clouds before it hits the river below. The long narrow ribbon of water falls so far that it is five hundred feet wide at its base in the wet season, and is mere vapor in the dry. The adventurous can get to the falls by taking guided boat trips up the river in the rainy season, then hiking overland for an hour to reach the falls. The falls can be seen only from the air during the dry season, and hopefully from a sound airplane—for it is still a long walk out of the rainforest from Angel Falls.

248 Bottom
*Canaima Falls is another
spectacular waterfall within the
National Park, wholly different in
its rushing blunt force from the
silken, narrow thread of the
graceful Angel Falls.*

249
*Brought to world-wide notice by
aviator Jimmy Angel in 1935, Angel
Falls is a ribbon of water that drops
2,648 feet, hits some rocks, and spills
nearly a thousand more feet to the
river below.*

250–251
Winding for thousands of miles through the South American rainforests, tributaries of the Amazon like this carry not only water but life to the plants and animals of the forest.

250 Bottom left
Scores of hummingbird species live in southeastern Venezuela. Hummingbirds favor red flowers and visit them from the forest floor to the canopy.

THE LUNGS OF THE PLANET

THE AMAZON RIVER AND ITS RAINFOREST

Birds call out incessantly as you hack your way up the hillside with a machete. The thick tangle of vegetation hides every imaginable form of life. Insects buzz around your face and crawl up your legs, while a howler monkey calls from a nearby treetop. Light from the sun penetrates the forest at a sharp angle, throwing harsh black shadows on the surrounding jungle. Suddenly you arrive at a clearing and see one of the most beautiful sights you have ever experienced: a young, agile jaguar in its native habitat. The trip was worth the discomfort, for in a split second you were able to come face to face with one of nature's most elusive creatures. Scrambling back down the hillside to the river, you note the wide, delightfully muddy expanse of slowly moving water and know exactly why it was that you came to this place at this moment of time.

The mighty Amazon River is the principal river of the South American continent and a major factor in the ecosystem of planet earth. All of the statistics regarding this river are huge. It rivals the Nile as the longest on earth, and it contains more water than the Nile, the Mississippi and the Yangtze rivers combined. The Amazon is 3,900 miles long, the second longest river in the world. The Amazon River and its more than one thousand tributaries form the world's largest drainage basin, taking in nearly two fifths of the entire continent of South America and two thirds of all the world's flowing river water and fresh water.

The Amazon rises in the Andes Mountains as two rivers, the Maranon and Ucayali. At their junction in the lowlands of northeastern Peru the river begins to be called the Amazon. The river flows eastward through the rainforests of Brazil, becoming larger at its junctions with the Japura, Jurua, Negro, Madeira, Tapajos, Purus, Teodoro, and Xingu rivers. As one travels on the Amazon or one of its tributaries, one cannot help but be impressed with the vast expanses of verdant tree-covered lands teeming with hidden wildlife. Colorful birds fly from branch to branch in the trees overhead, while monkeys leap and swing with agile grace. Moisture rises from the forest floor, and combines with the slanting rays of the sun to create a misty, mysterious environment. The visitor to this world can seem lost in time and space. How long ago did the Amazon receive its name? The year 1541 does not seem that long ago in the heart of the rainforest, the year when early Spanish explorer Francisco de Orellana was attacked by native people who wore clothing that reminded

250 Bottom right
This little fellow is a coati, sometimes called a coatimundi. A mammal with a slim body, long tail and long, flexible snout, the coati lives in the jungles of Mexico and South America. The coati likes to eat birds and fruit.

251
The Pantanal is an area of savanna, marsh, and forest in southern Brazil, near the border of Bolivia and Paraguay. It is located outside the Amazon drainage south of the Mato Grosso Plateau, in the Paraguay River basin. It, too is a haven for wildlife.

him of the female warriors in Greek legend. Orellana was the first European to travel the length of the river, and the name Amazon has been applied to the river ever since.

A large river like the Amazon has carried away so much of the surrounding land over the centuries that it creates a broad, low-lying plain for itself composed of thick sand and mud. The gradient down which the Amazon flows has been reduced to three inches per mile. In fact, the point at which the river crosses the Peruvian border with Brazil is only two hundred feet higher in elevation than that of the river at its mouth. And it is for this reason that, despite the massive discharge of the river, the flow of the water is very slow, about one and a half miles per hour in the dry season, and three miles per hour in the rainy season. Ocean tides overpower the current at the mouth of the river, creating a bore, or wall, of water fifteen feet tall. Effects of the tides can be seen up to six hundred miles inland.

The Amazon basin receives more than sixty inches of rain

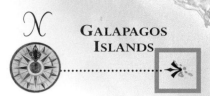

PACIFIC OCEAN

GULF OF
MEXICO

CARIBBEAN SEA

ATLANTIC OCEAN

GALAPAGOS
ISLANDS

SOUTH
AMERICA

"A LITTLE WORLD WITHIN ITSELF"

THE GALAPAGOS ISLANDS

Charles Darwin called the Galapagos "a little world within itself" and thought that they brought people closer to "that mystery of mysteries, the first appearance of new beings on this earth." Darwin visited the islands in 1835 when he was a naturalist on the British warship H.M.S. Beagle. The ship was making a five-year voyage around the world to draft charts for the Royal Navy. Since human interaction with the Galapagos had been minimal and since the islands were separated by so many miles from the mainland, they were a veritable laboratory of interesting and unknown species.

Darwin was fascinated by the closely related species he found on the individual islands, similar to one another yet different enough to cause him to ask questions. For instance, the giant tortoise had evolved into fourteen different forms on the various islands of the archipelago. Why? And why didn't giant tortoises of this type live on the mainland? The questions and specimens he collected lingered in Darwin's mind for twenty years as he pieced together his theory of natural selection. When Darwin published *The Origin of Species* in 1859, he shocked the world. At the same time, he gave scientists an entirely new way to look at biology and animal development.

The Galapagos Islands are located on the equator in the Pacific Ocean, about 650 miles west of the South American mainland. There are 19 islands in the group, with 107 islets and large rock outcroppings. Five islands are inhabited by humans, with a total population of about fifteen thousand. The islands are volcanic in origin and several volcanoes to the west of the Galapagos are still active.

The Galapagos are a part of the Republic of Ecuador and were annexed by that nation in 1832. The islands were discovered in 1535 by Tomas de Belanga and were later used by pirates as resupply points, particularly for water and for live giant tortoises, which could be stored aboard ship and killed for food as needed.

Today, the islands are managed by the Galapagos National Park Service, which has set up sixty specific sites for visitors to see. The remainder of the islands outside these sites are off limits to visitors. This is because the animals of the Galapagos are not used to human beings and have no fear of man. The animals are being studied and protected, for the islands today are every bit as useful as a biological laboratory as they were in Darwin's day. The Charles Darwin Research Station, managed by the international Charles Darwin Foundation, is located at Puerto Ayora on Santa Cruz Island. The Station has a successful program of captive breeding of giant tortoises and land iguanas. The Galapagos park was established in 1959, when 97 percent of the islands were set aside as a preserve. A 1986 law protected the waters around the archipelago, and in 1998 Ecuador enacted stiffer laws to protect the islands, especially from introduced plant and animal species. The islands became a World Heritage Site in 1978.

All tours, whether they are cruises or day trips, have to be accompanied by a guide licensed by the Galapagos National Park Service. Animals cannot be touched and visitors must stay on the trails. While traveling from site to site by boat,

258–259
Sometimes called the Enchanted
Isles because they tend to appear
and disappear in the heavy mist,
the Galapagos has nineteen main

islands. The island of Bartholome
is seen here in the foreground
with San Salvador behind.
Bartholome is a young volcanic
island.

259 Bottom right
Highland forest growth on Santa
Cruz includes scallcia, or tree
daisies, and bromeliads and
orchids. The crater seen here was

caused by the sinking of limestone
rocks undermined by caverns. The
island's highest point is 2,835 feet
above the ocean. The coasts of
Santa Cruz feature giant cactus.

258
The island of Española features
colorful marine iguanas, this one
seen sporting a lava lizard on his
head. Espaniola is a small island
in the southwest corner of the
archipelago.

259 Bottom left
Wild tortoises live amid the
vegetation in the highlands of
Santa Cruz. The entire chain of
islands takes its name from them,
Galapagos, the Spanish word for
tortoise.

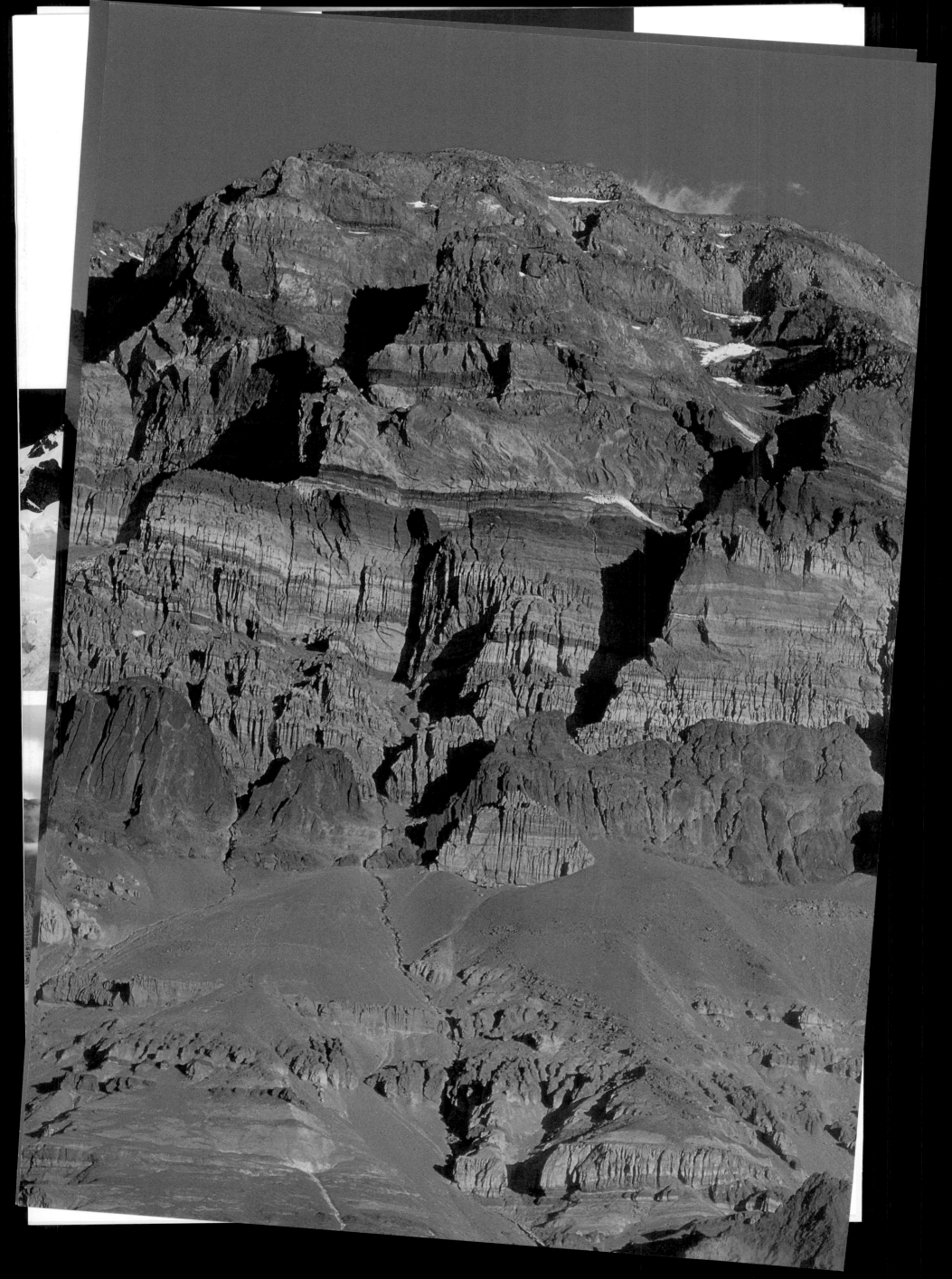

SOUTH
AMERICA

AFRICA

PACIFIC
OCEAN

ATLANTIC
OCEAN

INDIAN
OCEAN

AUSTRALIA

PACIFIC
OCEAN

N

ANTARCTICA

272–273
Antarctica is a land of haunting
beauty, as evidenced in this scene
of land and icebergs taken with the
sun low in the sky. Antarctica lies
beneath an ice sheet nearly as large

as the continental United States,
and stores nearly seven million
cubic miles of ice. The ice is
extremely heavy, and if it were to
melt, Antarctica, relieved of its
burden, would rise 2,800 feet.

CONTINENT of MYSTERY

ANTARCTICA

272
The weird, ghostly effects of the
Aurora Australis are captured
here, with the stars of the
southern sky shining through.
This view was taken at the South
Pole, and the buildings in the
background (distorted by the

wide-angle lens) are those of the
Amundsen-Scott base, a
scientific station maintained
since 1956 by the United States
on the Pole itself. The South
Pole was first reached by
Norwegian Roald Amundsen on
December 14, 1911.

273 Top right
A rocky island protrudes through the Antarctic ice in this beautiful scene. The thickness of the ice varies, but has been measured at several thousand feet. The greatest measurement ever taken was on the Hollick Kenyon Plateau, where the ice measured 14,200 feet thick.

273 Top left
Paulet Island, located near the northern tip of the Antarctic Peninsula, is seen in this view. Antarctica looks like a large round, slightly off-center wheel when seen on a map and the Antarctic Peninsula is the long, pinwheel-like arm that stretches toward Tierra del Fuego. Paulet is continuously locked in the embrace of the coastal ice pack.

Barren. Cold. Desolate. Lonely. Antarctica is all of this and more. But for all the negative things that can be said about Antarctica, much that is positive can also be discovered. The polar regions held a peculiar fascination for explorers in the nineteenth and twentieth centuries. Carsten Borchgrevink, Sir Ernest Shackleton, Robert F. Scott, Roald Amundsen, Robert Byrd, and others brought attention to a region that had been long misunderstood. These men came to love Antarctica for what it was, not for what they wanted it to be. It was the last landmass to be explored, and it remains the continent about which the least is known.

It may be barren, cold and desolate but Antarctica is also beautiful. Cool shades of blue, purple, gray, and white pervade the scene, punctuated by brown and tan rocks that rise above the snow and ice revealing Antarctica's underlying layer of earth. Huge ice floes float around its edges, while vast plateaus of slick blue ice stretch toward the horizon, reflecting glimmers of light from the sun lying low to the horizon. The brown peaks of rocky mountains can be seen protruding above the snow. Icebergs are calved from the shore, floating blue and white giants, and move outward away from the shores of the continent. Schools of fish and mammals such as whales and seals search for fish in the waters. Birds hover in the air, waiting to sight a fish and dive after it. Penguins poise on the edge of the ice floes looking for a meal, or move about in colonies.

In 1773 Captain James Cook was the first to observe and chart the edges of Antarctica. Individual expeditions of several countries, most notably Norway and Great Britain, attempted to reach the South Pole, but it was not until 1957 and the International Geophysical Year that intensive scientific research was begun there. In that year twelve nations set up research stations on and around the continent, including one at the Pole. At the end of the Geophysical Year, the Scientific Committee on Antarctic Research was organized to coordinate future efforts. The 1959 treaty, signed by twelve nations and now numbering twenty-six, ensures that Antarctica will not be used for military purposes or nuclear weapons or waste storage,

281
An iceberg floating in Paradise Bay seems to glow from within, shining against the dark magenta background of the rocky Antarctic coast. Most of the continent is relatively featureless, level or gently tilted. Antarctica is the world's highest continent on average, with elevations over a mile above sea level.

282–283
A flock of Emperor penguins is seen here against the backdrop of Antarctica's ice floes and snow-covered mountains. Of the world's eighteen species of penguins, the Emperor is the largest, standing forty-eight inches tall. Penguins are flightless birds peculiarly adapted to living in Antarctic conditions.

280 Top left
The penguin is a flightless bird adept at swimming in the ocean and catching fish. Here, Adelie penguins line up to dive for a meal. Slow on land, penguins are fast in the water, reaching speeds of twenty-two miles an hour.

280 Top right
The Ellsworth Mountains are the highest peaks in Antarctica, averaging 16,000 feet. The Vinson massif in this range is the tallest at 16,860 feet. The continental shelf—the submerged part of the continent—is the world's deepest, probably because it is weighted down so much by the ice.

280 Bottom
Two of Antarctica's adult Emperor penguins frame a fledgling. The Emperor, unlike other penguin species, breeds in winter when the weather is at its worst and Antarctica is shrouded in almost continuous darkness.

ANTARCTICA

seals, nine species of whales, seven species of penguins, and forty other species of birds, including petrels, live in the coastal waters. The larger animals and fish depend upon a food chain that begins with phytoplankton, which are free floating plants. These include about one hundred species of diatoms, single-celled microscopic organisms. Phytoplankton are consumed by such zooplankton as copepods and euphausiids, particularly billions of tiny shrimp called krill. Whales and fish consume krill, and the fish are in turn eaten by seals, penguins, and petrels. This fragile food chain allows animals to live in this hostile environment. But adaptation is necessary to survival. The fish in the Antarctic waters, for instance, have a substance in their blood somewhat like antifreeze. Studies of these fish are leading to scientific breakthroughs in several areas. Although the number of individual species is low, the animals themselves are numerous. Cape Adare, for instance, is home to approximately 260,000 pairs of Adelie penguins, the largest colony anywhere of this species. Whales include such species as the bottlenose, humpback, and finback. These make up the bulk of the population. Others, like sperm and blue whales, also inhabit these waters.

During the twentieth century, men and women have come to this inhospitable land to learn and to study, to marvel and admire nature's wonders. Near the Mount Erebus volcano, research stations for the United States and New Zealand are built in the area of Robert Scott's ill-fated camp of 1912. The unusual conditions of the south polar region have led to special adaptations of wildlife. Conservation will be very important here because the fragile web of life will not be able to withstand pollution, which has already reached it. Antarctica today remains the most mysterious and unexplored continent, filled with many inspiring little-seen wonders.

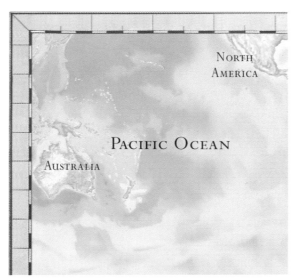

284 Top
Three hundred miles west of the Society Islands lies Aitutaki Atoll in the Cook Island chain. The atoll forms a large triangle with this island, Motu Tapuaetai, at the southeast corner.

284 Bottom
The gorgeous coastline of one of the eight major Hawaiian Islands is seen in this view. Moderate year-round temperatures of about 75 degrees Fahrenheit create a delightful tropical climate.

THE SOUTH SEAS

INTRODUCTION

To some, the region is known as the Pacific Islands, to others, Oceania. Whatever it is called, it is a vast area of the earth, taking up at least one third of the planet. Unlike the large continents, the region is more diverse, more far flung, and perhaps even more exotic than anywhere else. More than thirty thousand islands are scattered over the map of the Pacific Ocean, some so tiny and barren they cannot support life, some very large, like New Guinea or New Zealand, and one, Australia, a continent.

The largest clusters of the Pacific Islands are located in the southwestern portion of the ocean, where the seafloor features high volcanic ridges and deep troughs. It is the tops of these ridges, or individual peaks along them, that rise above the level of the ocean and appear as islands. Most of these peaks are active volcanoes, some continually venting steam and red hot lava onto the surrounding land and water. Some rise high above the water and are viewed as the mountainous regions of islands. Some lurk below the surface with coral reefs attached around the lip of their calderas forming atolls. Whatever their type or status, most are thickly forested with exotic plants that evolved independently of those in other parts of the world. Lush greenery, gorgeous waterfalls, tropical climates, and cooling thunderstorms mark these tiny dots in a vast, deep ocean.

Oceania is divided by geographers into three regions: Melanesia, Micronesia, and Polynesia. This division is based not only on the geographical types of islands but also on the cultural groups of people who settled them. Micronesia, which means little islands, contains twenty-five hundred islands, including the Caroline, Gilbert, Mariana, and Marshall Islands, as well as Palau and Wake. Most of the islands in the group are low coral atolls. Melanesia, which means black islands, includes New Guinea, Fiji, New Caledonia, the New Hebrides, and the Solomon Islands. Most of these islands are the tops of volcanoes and some are ridges of folded sediments. Polynesia means many islands and includes thousands of islands inside a huge triangle of ocean five thousand miles on a side. Most of Polynesia consists of islands that are the tops of huge volcanoes rising from the ocean floor.

The key to the beauty, history, and exotic nature of the islands is the ocean. Sometimes calm, sometimes whipped to a frenzy by wind and rain, the Pacific Ocean has many moods. It can be a lonely place and perhaps that is why the sight of the green islands is so welcoming. Gorgeous, multi-colored fish swim beneath the surface offshore, flitting among living coral reefs. The ocean's tides provide the rhythm of life in the islands.

The Pacific Ocean is the largest, deepest and oldest ocean basin on the planet. The seafloor contains rocks dated at about two hundred million years. The region has been called the Ring of Fire because of all the volcanoes around its

285 Bottom left
The Great Barrier Reef Marine Park, off Queensland, Australia, is seen in this photo of Whitsunday Island, north of Mackey. Beneath the waters of this incredible place lie coral reaches of unimaginably vast proportions, teeming with fish and marine life.

285 Bottom right
The island of Kauai, in the Hawaiian chain, rises dramatically from the Pacific Ocean. It is composed of a single eroded volcanic shield, with steep cliffs and inland canyons. Kauai's Na Pali Coast is so rugged that a road could not be built across it.

284–285
In this excellent aerial view of a coral atoll, the viewer can easily see the light blue waters of the protected lagoon and the forested land of several islands above sea level. The photograph is of Nokanhoui Atoll, one and a half miles in diameter and lying east of the Ile de Pins near New Caledonia. The atoll is about five hundred miles east of Australia.

286–287
This fantastic landscape is
located in Western Australia's
Purnululu National Park near
Kimberley. It is an excellent
example of the varied
topography of the region,
created from sandstone about
four hundred million years ago.
Sediment was washed into a rift
valley and created an underwater
reef. Rising above the waves, the
reef was later eroded by wind
into these strange beehive
shapes, called bungle bungles,
bordering deep canyons.

286 Bottom
This serenely beautiful scene of
snow-capped majesty seems
out of place in the South Seas,
but illustrates the diversity of
geographical features in the
region. Ngauruhoe Volcano
stands at 7,515 feet high above
sea level in New Zealand's
Tongariro National Park.
Located in the south central
area of New Zealand's north
island, Tongariro also includes
the 9,175-foot mountain called
Ruapehu.

287 Top left
Shark Bay in Western Australia is
on display in this view from
space. Sedimentary rock here was
created from algal deposits
millions of years ago to form
mounds as high as three feet
that look much like coral. Shoals
formed from these deposits
border tidal passes in the
Disappointment Reach shallows,
regulating water entering the bay.
Despite its name, Shark Bay is
home to dolphins rather than
sharks.

287 Top right
This view from space of
southeastern Australia gives an
idea of the vast array of
topographical features in the land
"Down Under." Lakes located in
the desert, such as Lake Eyre in
the lower right hand corner,
contrast with the coastal river
systems, the green of the
vegetation, and the blue of the
ocean at the top of the picture.
The city of Melbourne is located
at the top left of the photo, on
Port Philip Bay.

287 Bottom
A vent of the Yasur Volcano erupts with molten lava in this view of Tana Island in Vanuatu, formerly the New Hebrides chain of islands. At 3,420 feet tall, the peak of Yasur is the highest point on Tana, an island twenty-five miles long and fifteen miles wide at its widest point.

INTRODUCTION

rim. Molten rock wells up from beneath the earth's crust to the joints between these plates. As the hot rock cools beneath the ocean floor, it adds new crust to the plates on both sides, thus putting tremendous pressure on the plates. As the plates press against each other they have to give at some point, and finally they buckle, creating undersea mountains and trenches, called subduction zones. In these deep trenches some of the crust is carried back down to the mantle. As a result of these forces, the Ring of Fire has many active volcanoes, some underwater. Found on the landward side of oceanic trenches, volcanoes spew forth lava and add land to islands and continent rims.

It is the lava that made the soil of the islands such rich, fertile ground for lush vegetation. Many of the islands are overgrown with tremendous numbers of plants, some endemic and others imported by humans. The cool green retreats are home to many types of birds and animals, although most animal life was imported as well. Walking down a jungle trail one might encounter coconut trees, orchids, and other tropical flowers and exotic birds and wild pigs. Most of the volcanic islands have high central mountains and are small enough for the entire sweep of the land to be seen from one vantage point. It is from places like these that the true beauty of the oceanic islands is easiest to see. These emerald gems, with waterfalls and streams dropping off on all sides headed for the sea, present a lovely counterpoint to the strong blue of the ocean and the white of the surrounding beaches.

The story of the habitation of these islands is as extraordinary as their beauty. The Pacific Islands were settled by intrepid explorer-colonists that ventured forth in large outrigger canoes from their homelands in Asia. One by one the islands were settled, beginning over 1,700 years ago and continuing until the furthest reaches of the Pacific had been colonized. The double-hulled canoes of these intrepid people contained taro and sweet potato roots, seeds and saplings of coconut and banana trees, sugar cane, medicinal plants, pigs, dogs, and chickens. Europeans did not know about the Pacific Ocean and its vastness until explorers like Ferdinand

Magellan sailed through the region in the early sixteenth century. Later, a small number of traders, entrepreneurs and missionaries from many nations came to inhabit the islands alongside the indigenous people. These populations swelled, especially in the case of the British settlers of Australia and New Zealand.

Today, the Pacific Islands range from independent nations to protectorates to possessions of European, American, and Asian nations. Their incomparable beauty, tropical weather, fascinating wildlife, and interesting heritage make them dream destinations for people in many parts of the world.

288–289
They rise from the desert sands 155 miles north of Perth in Western Australia, forming a haunting, unforgettable sight in Nambung National Park. Limestone spires formed by sedimentary layering and wind erosion over at least thirty thousand years, they stand as monuments to the more unusual beauties created by nature.

288 Bottom left
Ghostly visions sometimes make the Pinnacles seem like they were carved by the hand of man. A walk through the Pinnacles is a contemplative experience.

288 Bottom right
Like desert army tents, the Pinnacles sometimes seem carefully ordered despite the randomness of nature. The Australian deserts are most ancient and have changed little over time.

289 Top
Only the human being in the scene can give a sense of the scale of these eroded spires at Pinnacles, which look like pebbles cast on a giant's beach.

289 Bottom
The sharply spiked pinnacles are particularly striking when the sun is low in the sky and the light shines at an angle, throwing dark shadows on the flat sandy desert.

INDIAN OCEAN

CORAL SEA

WESTERN AUSTRALIA

AUSTRALIA

PACIFIC OCEAN

HAUNTING SPIRES

AUSTRALIA'S PINNACLES DESERT

Light winds blow the silent ochre sands around tall, strangely shaped spires protruding from the dunes. These randomly placed towers cast dark shadows on the ground while high cirrus clouds turn orange in the fading light of day. These are the Pinnacles, strange and haunting landmarks on the coast of Western Australia.

Just a 155-mile excursion north from the city of Perth through Australia's Outback, the Pinnacles Desert is reached by driving through pine plantations, bushlands, and, in season, an incredible profusion of wildflowers. Standing in Nambung National Park, more than a thousand acres of mysterious looking limestone spires dot the flat, sandy landscape, looking like the stumps of a clearcut forest.

These are the Pinnacles, bony fingers of rock pointing straight toward the sky, while the wavy patterns of endlessly shifting yellow and ochre sand swirl about their bases. They range in size from mere stony twiglike rocks to columns nearly sixteen feet tall. The Pinnacles are striking because their hard forms rise so brusquely from the soft sand around them. Geologists estimate that they are about thirty thousand years old. It is thought that a soft sandstone exterior was eroded away by wind and drifting sand over a period of thousands of years, leaving the more durable limestone core exposed. Some scientists have also theorized that the Pinnacles may be the fossilized roots of ancient trees. Whatever their history, they present a startling spectacle for visitors to the area.

The Pinnacles are located in Western Australia, the largest state of the island continent. Western Australia occupies the entire western third of the country and is composed almost entirely of desert. Although it receives about ten inches of precipitation annually, more rainfall than many desert regions in the world, Western Australia is still a thirsty place nearly devoid of moisture. Water arrives suddenly with monsoons or tropical cyclones, then evaporates quickly. Thus the forces of wind and sand are able to perform their work swiftly under the hot Outback sun.

Although the Pinnacles Desert exists in this harsh environment, it is not really a desert at all but a series of flat coastal sand dunes. The Pinnacles themselves can be seen from the ocean and early explorers at first thought them to be the spires of a town. They do, in fact, bear a strong resemblance to traditional West African houses.

Today, Nambung National Park includes sixteen miles of coastline along the Indian Ocean, with sugar white sand beaches and incredible vistas of the rolling blue waters. In addition to geological wonders like the Pinnacles, Nambung also harbors about one hundred colorful bird species and many kinds of animals. Western Australia is famed for its wildflowers and the jarrah forests have more than three thousand gorgeous species that thrive between August and November. Kangaroos live in the park and can be seen in the vicinity of the Pinnacles.

These natural wonders, found along the west coast of Australia's most rugged deserts, are an eerie and beautiful example of desert scenery at its most spectacular. Despite these other attractions, it is the Pinnacles that draw visitors to the area. The Pinnacles represent the way in which shapes and natural forms please and inspire the human mind. A few of these simple and unusual features might be interesting, but one thousand acres of the gray stone towers, each unique but all part of the random whole, make Pinnacles a true natural wonder.

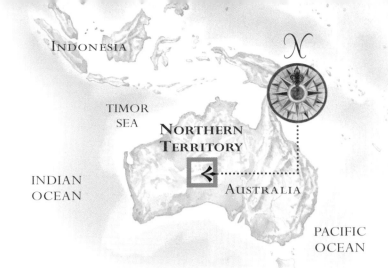

LANDMARK ON THE DREAMING TRAIL

A Y E R S R O C K

Nearly every evening of the year they gather. Tourists from all corners of the world stand in awe at a vantage point about three miles from a huge sandstone edifice, watching and waiting. As the sun sets and changes color, the rock does too, producing a strong and satisfied reaction from the crowd. A huge orange rock between flat desert and violet sky turns red and seems to glow in the fading light of day.

Ayers Rock is located in the Northern Territory of Australia and is perhaps the nation's best-known landmark. Known more frequently today as Uluru, the Aboriginal name for the site, Ayers Rock is a giant monolith that towers 1,141 feet over the surrounding Mulga Plains, which reach an altitude of about 1500 feet. The rock measures 2.2 miles long and over 5.5 miles in circumference. And that is just the part that shows! It is believed that Ayers Rock is the largest single rock in the world. At least two thirds of it is buried beneath the desert. In comparison, Devil's Tower in the United States, a similar feature that springs suddenly from the surrounding landscape, is 865 feet tall with a flat, mesa-like top covering just one and a half acres.

Ayers Rock is within relatively easy reach of the city of Alice Springs and is part of Uluru-Kata Tjuta National Park, administered by Australia's National Parks and Wildlife Service in conjunction with the local Aboriginal tribal councils. The park was created in 1985 and designated a World Heritage Site by the United Nations in 1987, one of the few that contain cultural and natural treasures that are equally unique.

The sheer, orange sandstone walls of Ayers Rock dwarf casuarina trees growing at its base. The rock is colored through oxidation of the iron components of the minerals in the sandstone. It is believed that the processes which created Ayers Rock as it exists today began five hundred million years ago when sediment began accumulating on the bottom of an ancient seabed. About three hundred million years ago the waters receded and buckling caused the sedimentary sandstone layers to tilt on end. By forty million years ago, wind and sand completed the process, sculpting and molding the part of the sandstone left above the surface.

Europeans discovered Uluru in 1872 when Ernest Giles was exploring the region. Beginning in the mid-twentieth

290

A close view of Ayers Rock reveals the rough texture of the rock surface and the vegetation that finds a niche and patches of dirt to begin growth. Five hundred million years of sedimentary buildup has combined with millions of years of erosion, first by water, then by wind, to create this incredible work of nature.

290-291
Located near the geographical center of Australia, in the southwestern corner of the Northern Territory, Ayers Rock looks as though it was placed upon the desert sands by the hands of a giant. Looks can be deceiving, however, for only the tip of the monolith protrudes above the surface, while it extends perhaps as much as three thousand feet below.

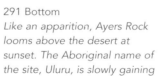

291 Bottom
Like an apparition, Ayers Rock looms above the desert at sunset. The Aboriginal name of the site, Uluru, is slowly gaining acceptance. Unfortunately most visitors ignore Aboriginal wishes that the 2.2-mile long, 1,141 foot tall monolith not be climbed.

INDIAN OCEAN

AUSTRALIA

GREAT BARRIER REEF

N

PACIFIC OCEAN

THE ULTIMATE WORK OF NATURE

T H E   G R E A T   B A R R I E R   R E E F

294
The Swain Reefs, located at the southern end of the Great Barrier Reef, are seen in this view from space. Australia's Great Barrier Reef runs for over 1,200 miles off the northeast coast of the island continent. The Great Barrier Reef is composed of thousands of individual reefs.

294-295
The Great Barrier Reef is a hazard to navigation, but at the same time forms a protective breakwater that actually shelters ships sailing parallel to it. The depth of waters along the reefs is very shallow, ranging from less than three feet to about 130 feet, while the offshore depths sink quickly to three thousand feet or more.

295 Top left
This view shows Heron Island, one of over six hundred islands located along the Great Barrier Reef. There are two types of islands found here: low-lying reef islands, and larger islands formed by the tops of mountains and volcanoes.

295 Top right
Many of the Whitsunday Islands, numbering over seventy, are part of the Great Barrier Reef Marine Park. In the north, the reef is virtually continuous and lies as little as ten miles from shore.

Hundreds of colorful fish glide past your facemask as bubbles escape from your mouthpiece. All about you in the warm water, colorful coral can be seen forming tiny towers, minarets, and spikes as fragile as china. In and among the coral, predators wait for a juicy meal to swim by, and swim they do, by the million. Light flashes off the fins and tails of incredibly hued fish. You can observe an entire ecosystem with the sweep of your eye, from the bottom to the top of the food chain. You are swimming in the Great Barrier Reef, the largest structure on earth made by living creatures.

The Great Barrier Reef is the world's largest mass of coral, extending for 1,260 miles in the Coral Sea off Australia's Queensland coast. Coral is produced by the accumulation of tube-shaped skeletons of billions of marine polyps of the phylum Cnidaria and algal plants called nullipores all held together by a limestone substrate. Polyps are animals that resemble tiny sea anemones. As coral runs through its natural life cycle and dies, reefs continue to grow as colorful new animals build on the white skeletons of the old. In order to survive, coral polyps need shallow salty water that is clear and warm, found mostly in tropical seas. That is why coral reefs generally form only between 30 degrees north and 30 degrees south of the equator. Depending upon condition—and coral reefs are very dependent upon exact conditions—a reef can grow anywhere from less than half an inch per year to forty inches per year. Over time, sand can accumulate on the side of the reef facing the shore, and some of the reefs become low islands.

Barrier reefs are broad coral platforms separated from coasts by wide channels. The Great Barrier Reef is actually a chain of over 2,500 reefs extending from a point opposite Mackay, Queensland into the Torres Strait that separates Australia from New Guinea. The shallow channel varies from three miles wide on the north to one hundred miles wide on the south and is dotted with islets and atolls, making it very difficult to navigate. The reef acts a protective barrier to the Australian coast.

296-297
The spectacular underwater world of the Great Barrier Reef includes incredibly unusual, colorful, and beautiful organisms.

For instance, these large yellow sea fans grow in a narrow channel of water, where their branches can capture sea creatures that swim by.

296 Bottom left
Countless colorful and exotic sea creatures live on the Great Barrier Reef. The coral deposits of the area are incredibly beautiful but also quite fragile.

296 Bottom right
A giant sponge is another denizen of the underwater world of the coral reef. Simple invertebrates, sponges have existed for least six hundred million years.

297 Below and bottom
Sea fans are composed of millions of little sea polyps. Each secretes calcium carbonate from the bottom half of the stalk of the animal, forming little cups into which the polyps can withdraw for protection. At night tiny tentacles extend from each polyp's cup to seize animal plankton that wash against the fan.

Queensland occupies in the northeast corner of Australia and is known as the Great Barrier Reef State. Cairns is the city that provides visitors with many of the diving charters, fishing trips, glass-bottom boat excursions and longer boat trips to the reef. After leaving Cairns there is more to the Great Barrier Reef than the world beneath the water. There are more than six hundred islands around the reef and on it, including excellent bases from which to scuba dive or snorkel, such as Lady Elliot Island and Heron Island. Nearly all of the seventy or more Whitsunday Islands are national park areas, and have forests of eucalyptus and acacia trees as well as hoop pines. In these forests live at least 156 species of birds and animals like wallabies, possums and goannas. Magnetic Island even has a sanctuary for koalas. Hitchinbrook Island is Australia's largest island national park, and is a relatively undisturbed area of rainforest with mangrove swamps and a mountainous region.

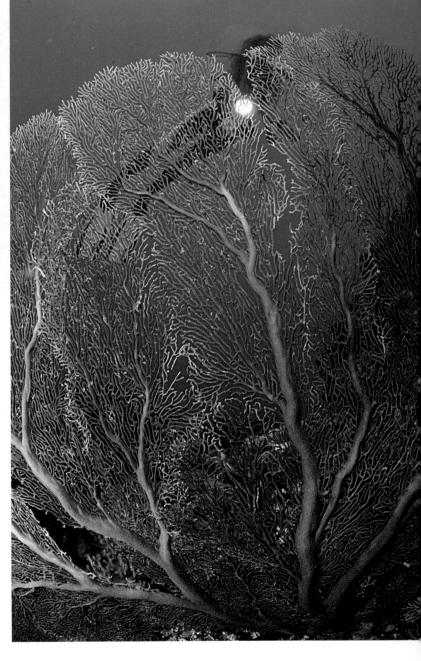

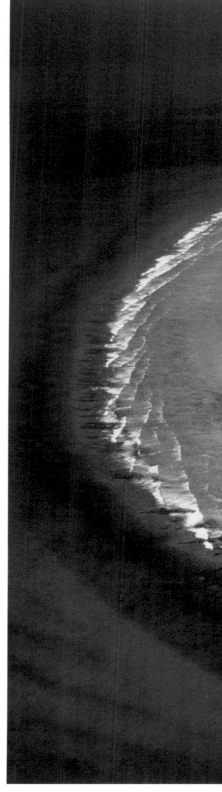

298 Top left
A huge manta ray, or devil-fish, swims off the reef, accompanied by pilotfish. Manta rays are the world's largest form of ray, reaching a twenty-foot "wingspan" and weighing up to 1,500 pounds. They eat plankton and small fish that they strain out of the water with their gaping mouths.

298 Top right
Black-tip sharks cruise the waters off the reef looking for a meal. Always hungry, sharks tend to congregate off the reefs in the deeper waters, but are known to enter the reef zones looking for food. Adult Black-tip sharks average five feet long. Sharks ranging in size and viciousness all the way up to the Great White Shark inhabit these tropical waters.

298 Bottom
A giant clam sits amid sea plants and atop hard coral. The giant clam is found on coral reefs in the warm waters of the Indian and Pacific Oceans. They are the largest living mollusks, weighing up to five hundred pounds and containing twenty pounds of flesh in the hatchet-shaped foot.

But the greatest attraction remains the world beneath the waves. The Great Barrier Reef is inhabited by a profusion of undersea life, including about two thousand species of fish and six species of turtles. Fish and other organisms include everything from clownfish to giant clams, rays, octopus, eels and parrot fish, from tiny gobies to huge whale sharks. Dangerous creatures inhabit the waters, such as both freshwater and salt water crocodiles, box jellyfish and sea wasps. Sea urchins, starfish, sea cucumbers, mollusks, and crustaceans are all here. Larger creatures like dolphins, humpback whales, sea turtles, and dugongs can also be seen.

The Great Barrier Reef Marine Park Act was passed in 1975 and established boundaries and a managing authority for the region. The Great Barrier Reef World Heritage Area covers an even larger area. The area needs protection from the thousands of tourists, recreational divers, boat enthusiasts,

snorkelers, and fishermen who want to see it for themselves. Coral needs very particular criteria to grow and can die very easily. As large as it is, the Great Barrier Reef is also quite fragile. Pollution, agitation of the water from boat wakes, destruction from people walking on the coral or breaking pieces off, and even human sweat can damage the ecosystem irretrievably. A huge threat is the sewage flowing from rapidly growing towns. The purpose of the Great Barrier Reef Marine Park Authority will be to study the extent of current damage and make recommendations for the future, particularly to control tourism and prevent future damage to the resource.

Damage comes in many forms, not all of it directly attributable to tourism. Coral reefs world-wide have been subject in recent years to bleaching, a condition of discoloration or loss of symbiotic zooxnthellae, round algae that photosynthesize and which are crucial to the existence of

298-299
One Tree Island sits atop the
Great Barrier Reef and emerges
from the water at the lower
center, while an expanse of coral
reef extends from it in a semi-

circular pattern. Within the arms of
the reef, the water is shallow and
light blue in color, while outside the
dramatic drop into the deep ocean is
signaled by the midnight blue of
the water.

299 Bottom
A school of snappers swarms in
the warm, clear waters of the
Great Barrier Reef. Parrotfish,
angelfish and butterfly fish are

some of the 1,500 to 2,000
species of fish to be found on the
reef, ranging in size from little
gobies to the gigantic whale shark
that grows up to fifty feet long.

healthy coral reefs. Some biologists think that bleaching is caused by waters that are too warm by a few degrees. Coral can grow in waters that cool to 61 degrees Fahrenheit at the lowest. The optimum temperature for coral is 78.8 degrees to 80.6 degrees Fahrenheit. Warmer water can kill coral and may cause bleaching. Another threat comes in the form of the crown-of-thorns starfish, an animal that eats living coral and is invading the Great Barrier Reef.

There is no doubt that the managers of such a vast natural area—especially a place so attractive to tourists and subject to so many natural threats—will have their hands full. But it is equally certain that the Great Barrier Reef is one of the major treasures of the earth, the most complex aquatic ecosystem with the greatest number and concentration of aquatic species on the globe. The Great Barrier Reef is a natural wonder without parallel.

INDIAN OCEAN

AUSTRALIA

PACIFIC OCEAN

NEW
ZEALAND

REACHING FOR THE SKY

NEW ZEALAND'S ALPS

A gigantic snow-capped range of blue mountains rises above the rounded green foothills and valleys. A white glacier flows down a large gap, turning gray as it flows and melts, passing through the green tree-covered plain to the lakes below. The images of the mountains are serenely reflected in the lakes at their bases. These are the Alps, but not those of Europe. Instead, these Alps are located on an island in the South Pacific close to Antarctica.

New Zealand can be described as a world in miniature.

On the two major islands of this spectacular nation in the South Pacific, nearly every type of geographical feature is present. The South Island includes fjords, snow-peaked mountains, glacial lakes, white water rivers, rolling farmlands and tropical forests. The North Island includes active volcanoes, erupting geysers, hot springs and sandy beaches. All of this adds up to a natural wonderland, but one of the most spectacular attractions of New Zealand is its mountains, known as the Alps because of their similarity to Europe's famed range. New Zealand's Alps are steep and rugged, support glaciers, and have their own unique ecosystem of flora.

Since New Zealand has been separated from all other land masses for eighty million years, many plants and animals have developed on the islands in isolation. Of some two thousand indigenous species, fifteen hundred are unique to New Zealand. Today, only about 15 percent of the land area of the nation is covered with native flora, as later human inhabitants imported many other animal and bird species.

The New Zealand Alps stetch for nearly the entire length of the South Island, over six hundred miles. Seventeen of the peaks in the range exceed ten thousand feet. The Alps were created by massive uplift, warping, and block faulting, then sculpted by glaciers much like their European counterparts. Many of the South Island's rivers rise in the Alps, including the longest, the Clutha River, which originates in two lakes, 48-square-mile Hawea and 75-square-mile Wanaka. The mountains act as a barrier for moisture-laden winds from the Tasman Sea which lies between Australia and New Zealand, causing the climate west of the mountains to be wet, while that to the east is dry. About eighty thousand New Zealanders make their living off the land from dairy farming, growing arable crops, tending orchards, and raising sheep, beef, and deer. Sheep raising is a particularly important part of the way of life in the Alps.

Mount Cook National Park is one of the premiere sites of the Alpine ecosystem of New Zealand. Viewed from the Hooker Valley, the mountains seem to leap straight up from the valley floor, without the gradual ascent of foothills. Mount Cook, called Aoraki by the Maoris, was 12,349 feet tall but was cut down to 12,316 feet in 1996 when it lost part of its apex as the result of a landslide. It is covered with snow all year, and it is still growing because of glacial rebound. As ice melts, weight is lifted off the mountain and it rises. Nearby Mount Tasman is the second tallest peak in

301 Top right
New Zealand's Alps are located on its South Island. Mount Cook, seen here, is the tallest of hundreds of beautiful peaks that form a spine along the nation's west coast, with over 140 peaks exceeding seven thousand feet.

301 Top left
The southern face of Mount Cook forms a perfect pyramidal shape on this clear day. The mountain recalls the indigenous Maori legend about Aoraki, the tallest and most important of four brothers turned to stone.

300
The new profile of Mount Cook is seen in this recent view, taken after a 1991 rock slide sliced thirty-three feet from the top. The forces of nature cut down the once 12,349 foot tall peak to 12,316 feet. Mount Cook was first scaled in 1891 by Englishman Edward Fitzgerald and was a favorite training ground for Sir Edmund Hillary in the late 1940s.

300-301
A ski mountaineer is treated to an incredible view near the summit of Von Bulow Peak in New Zealand, with the breathtaking summit of Mount Tasman behind. New Zealand's second tallest mountain, Mount Tasman suggests the incredible beauty of this island nation.

the range, and lies at the center of superlative mountain, forest, and lake scenery.

Many wildflower species grow only on the South Island of New Zealand, including the South Island edelweiss, the New Zealand buttercup, the Mount Cook lily with its pure white petals and yellow center, and the large mountain daisy. Lush tree ferns inhabit the forests of the foothills, displaying nearly every imaginable shade of green. In addition to uniquely beautiful flora, the New

Zealand Alps also include a bevy of interesting fauna. The kea, which inhabits the New Zealand Alps, is the only species of mountain parrot in the world.

Lakes bordering the eastern side of the Alps, like Lake Tekapo and Lake Pukaki, have incredibly blue waters—in fact they look artificial. This effect is natural and is caused by the runoff of water into the lakes from glaciers. The water carries fine particles of pulverized rock called rock flour from the glaciers down to the lakes. The rock flour stays suspended in the water and turns it an amazing bright blue. Glaciers send icebergs into the lakes surrounding the mountain area, and are actively carving land and even new lakes out of the countryside. As in other parts of the world, New Zealand's glaciers are shrinking and the effect of their scouring action can be observed quite well in their areas of retreat.

New Zealand's unique natural heritage makes it one of the world's most important geographical areas, exhibiting a multitude of wonders. Above them all, the beautiful Alps create an indelible impression of beauty and grandeur.

302
Lake Marion in Fjordland National Park is seen in this view of the very southwestern corner of New Zealand's South Island. The park is composed of spectacular and substantial peaks that surround glacier-fed lakes and fjords.

302-303
Mount Tasman sits astride the ridge between Westland and Mount Cook National Parks in this stunning sunset view. The massif is covered by a magnificent glacier, one of the largest to be seen anywhere outside of the polar regions.

303 Bottom left
Mount Cook and the Southern Alps are seen here. Glaciers cover at least 40 percent of Mount Cook. The valleys on the west side of the peak are rugged and steep, with deep gorges and thick vegetation below the tree line.

303 Bottom right
Mount Cook National Park includes far more than just mountain peaks and their glaciers. The park preserves spectacular Alpine areas of great natural beauty, with mountain meadows and lakes.

New Zealand's Alps

GOD'S AQUARIUM

THE PALAU ISLANDS

SOUTHERN
ASIA

PACIFIC
OCEAN

PALAU ISLANDS

INDONESIA

PAPUA
NEW
GUINEA

AUSTRALIA

304

The Rock Islands are a small portion of a much larger archipelago of the Palau group. Located at the crossroads of three of the world's major ocean currents, the waters off Palau are perfectly suited to tropical fish, coral, anemones and other sea plants. The crystal clear waters boast visibility of up to 150 feet.

304-305

The gorgeous sweep of Palau's Rock Islands and the Seventy Islands National Park can be seen in this view of the Pacific paradise. The region includes three hundred mushroom-shaped islets, little floating gardens peppered with white sand beaches. Beneath the waters lies the complex world of a coral reef.

The brush is thick and lush as you make your way along a nature trail and the scent of orchids is in the air as you approach the beach. Emerging from the forest, the sunlight seems too bright, the sand too white and pristine. Colorful birds fly by and part of the coral reef upon which the island rests can be seen just beneath the clear green waters. Overhead, the branches of leafy trees throw shade out over the beach, aided by the fronds of a palm tree high above. This is one of the incredible islands of the Pacific nation of Palau.

The Palau (pronounced beh-LAU) Islands lie about five hundred miles east of the Philippine Island of Mindanao in the western Pacific and about five hundred miles north of New Guinea. They are part of the Caroline group of islands in the area called Micronesia. The chain of 343 volcanic islands and a few coral atolls extends for about one hundred miles from north to south, about twenty miles from east to west, and is surrounded by a coral reef. Few of the islands are inhabited.

People first came to Palau about 1000 B.C., probably from Indonesia, New Guinea and Polynesia. They developed a unique culture with painted carvings, wooden storyboards and delicate woven items. The first European contact occurred in 1783 when Captain Henry Wilson of the English vessel Antelope was shipwrecked on Ulong. The Palauans helped Wilson repair his ship and return to his home country. The United States occupied the islands after World War II and administered them in trust for the United Nations. The Palauans formed a government in 1981 and became a self-governing nation in 1994. Today, the Republic of Palau is divided into 16 states with its capital at Koror and covers about 196 square miles of land. The nation has a developing agrarian economy and a growing tourism industry.

Palau's tropical climate and beautiful white sand beaches lure tourists. Palm and coconut trees line the shore and lean toward the ocean at seemingly impossible angles. Lush tropical gardens sport flowers such as orchids, while delicate petals of deep red and yellow accent blossoms. Biologically unique plants compose Palau's forests and fifty species of colorful tropical birds make their homes there. Palau highlights its traditional customs, crafts, and culture in public demonstrations at resort hotels and sponsors tours to traditional villages and ancient stone monolith sites. Recreational activities such as scuba diving, snorkeling, windsurfing, and kayaking are very appealing to

305 Bottom right
The fringing coral reef of Palau Atoll creates a ninety-mile lagoon. Palau is a coral atoll composed of 343 islands, the *largest being twenty-five-mile-long Babelthuay. Although the islands cover some 196 square miles of land, most of Palau is composed of reefs and water.*

305 Bottom left
The Kayangel Atoll lies to the north of the main Palau Atoll.

This aerial view shows the top of the sunken volcano's caldera and the areas colonized by coral.

310-311
This aerial view of Bora Bora clearly shows the outline of the encircling coral reef and the main, tree-shrouded islands. Taken looking westward, the picture shows Motu Piti Aau in

the lower right hand corner, with clouds shrouding the main island at the top right. The southern end of Bora Bora is where most of the island's popular resorts are clustered.

311 Top left
Raiatea is famous as the cultural heart of Polynesia. This sacred island has fewer beaches than Tahiti and Bora Bora but features a large protected lagoon.

311 Top right
An aerial view of Moorea, with Tahiti on the horizon in the distance, gives some idea of the lush green paradise. The island is actually part of the south rim of an ancient volcano.

in circumference. Otemanu is the largest remaining subaerial section of the volcano that formed the island. A coral reef encompasses the Island of Bora Bora, forming a lagoon with light blue water which is world famous for its beauty. White sands lead to perfect beaches fringed with palm and coconut trees.

In contrast, just twenty-five miles west of Bora Bora lies tiny Maupiti, one of the most beautiful and least visited islands in Polynesia. Placid lagoons, coral reefs and beaches provide a true escape. Likewise, the sister islands of Raiatea and Tahaa are today what Tahiti was many years ago. Tahaa is famous for its production of vanilla, an introduced plant of the orchid family, and for its nature preserve, which shelters the endangered marine turtles that thrive in its waters. Raiatea is considered to be the cultural heart of Polynesia, a sacred island where the early settlers who arrived from Hawaii first stopped.

"Getting away from it all" has new meaning on Fatu Hiva and Hiva Oa in the Marquesas. These gorgeous islands are farther from continental landfall than anywhere else on earth. The twelve islands of the group are located just south of the equator and are shrouded in constant cloud cover. Fatu Hiva is a volcanic rock six miles long, overgrown with lush vegetation that comes up to the water's edge, sheer cliff faces, rushing streams, waterfalls, and prominent rock pinnacles above the hidden valley of Hanavave village. Plants and flowers include orchids, spider lilies, ginger, ylang-ylang, jasmine, plumeria, and bougainvillea. Larger than Fatu Hiva is its neighbor, Hiva Oa, where the French painter Paul Gauguin is buried. Nearby Taiohae, the capital of the Marquesas, is blessed with steep mountains and valleys immortalized by the American author Herman Melville in his book *Typee*. It is the wettest of the Marquesas Islands, with nearly constant rainfall for a large

311 Bottom
Another aerial view of Bora Bora, looking eastward, shows the arm of the club-shaped peninsula called Motu Piti Aau along the top and the tip of Pointe Matira on Bora Bora at the bottom center. Matira Beach with its dazzling white sand and warm, shallow waters is a favorite spot for tourists.

FRENCH POLYNESIA

portion of the year. The Mouake viewpoint on Taiohae rises to 2,800 feet providing a stunning view of the bay, while an inland waterfall in the Hakaui Valley plunges 1,148 feet.

Beautiful white sand beaches line the northwest shore of little Tahuata Island that occupies only nineteen square miles of land. For such a small place, the island has a rich history. Tahuata was settled by about 300 A.D., was visited by the Spanish explorer Mendana in 1595, and James Cook in 1774. The French established a garrison there in 1842 and many historical buildings can be seen today amid the lush greenery. The island of Ua Huka is drier, lower in elevation, and less forested than other islands in the Marquesas and has goats and wild horses that graze on its upland plateaus. Ua Pou (the pillars) is one of the most spectacular islands in Polynesia. Huge peaks reach for the sky, straining four thousand feet upward from the shore.

In contrast to the Society and Marquesas Islands, the Tuamotus are a series of seventy-eight low-lying coral atolls. Geologically these are the oldest islands in Polynesia, the vestiges of ancient volcanoes that sank over the eons. Coral growth on the rims of these volcanoes keeps stretching for sunlight, so gigantic O-shaped rings of coral, known as atolls, have formed. Atolls rarely rise more than six feet above sea level and are covered with thick vegetation. The atolls of Tuamotu are remote and its people cater to tourist needs, collect copra, and search for seashells.

Some of these atolls are huge; for instance Rangiroa is forty-two miles across, and is the second largest atoll in the world. Fakarava, thirty-seven miles across, has an immense lagoon with several black pearl farms and many breeds of nesting birds. Lagoons are formed inside the ring of an atoll that provides habitat for creatures not at home in the open sea. Narrow openings let tidal waters spill into the lagoons, providing nutrients for coral and marine life. On Fakarava, a little town called Tetamanu is located near one of these openings where, when the tide is even, a virtual parade of marine life swims into and out of the lagoon.

When sea levels rise as a result of global warming, these atolls will all be drowned. These otherworldly areas are already suffering. Today they remain the ultimate escape to paradise—imagine sitting peacefully in a coral inlet watching colorful fish swim by—and it is crucial that they be preserved.

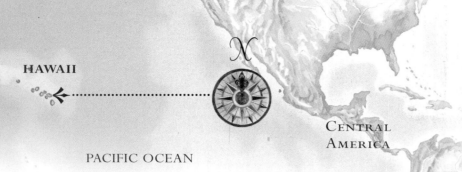

HAWAII

PACIFIC OCEAN

CENTRAL
AMERICA

312 Top
Extinct smaller volcanoes dot the landscape of the islands. Over the course of thousands of years, spores and seeds made their way to the fertile islands on winds and waves.

312 Center
Black volcanic rock created by ancient lava flows is seen in this photo of Oahu. This type of rock is plentiful on all the islands, as they were all created from volcanic activity.

312 Bottom
A wild, searing eruption of Mauna Loa volcano is seen here in 1984. At 13,677 feet Mauna Loa is the tallest active volcano in the world. During an eruption, molten rock, called magma, rises from the earth's mantle to a reservoir two miles below Mauna Loa's summit.

313
Mauna Loa's 1984 eruption spewed a river of lava down the mountain's northeastern face. Other major eruptions with large amounts of lava spill took place in 1942, 1949, 1950 and 1975.

LAND
FORGED
BY FIRE

HAWAII VOLCANOES NATIONAL PARK

Red, angry lava oozes up from out of the ground, while a sulfurous stench fills the air. Pink steam clouds rise from the black earth and seem to glow from within. Behind the smoke, a gorgeous display of orange fireworks sends traces of molten rock from the inner recesses of the earth arcing through the air. As the hot lava reaches the ocean, it spills in, hissing and fizzing as it hits the cooler water, exploding as it solidifies to create new land. For as the molten lava cools, the result of its spectacular display is the formation of terra firma. Billions of years ago this scenario was common all over the globe. Major portions of the earth were formed in just this way. Today, there are few places where volcanic activity like this can be witnessed first hand as it can on the island of Hawaii.

Hawaii is one of the Hawaiian Islands, a Polynesian group in the mid-Pacific composed of eight main islands and 124 islets, reefs and shoals. One of the fifty United States, Hawaii is a land of exceptional physical beauty. The islands rise up from the ocean, in the case of the peak of Mauna Kea, dramatically to over 13,700 feet. The Hawaiian Islands, like the others in the Pacific, are actually the tops of sunken volcanoes.

Hawaii is the largest of the island group, the youngest, the furthest south, and it is growing every day due to the accumulation of lava on its shores. Hawaii is located directly above one of the primary hot spots of the Pacific Ocean crust. Lava bubbling from the volcano called Kilauea flows into the ocean, killing trees, farm fields and everything else in its path while forming new land when it hits the cooling waters.

Kilauea is not like our mental picture of what a volcano should be: a tall mountain peak with a crater (caldera) at its summit, with hot red lava spilling down the sides. Instead of pouring from the top of the mountain, lava has found an easier escape route, through a vent on the lower side of Kilauea

The entire chain of the Hawaiian Islands can be seen in this view from space looking southward.

This view from space shows the entire island of Hawaii, highlighting its forested areas in *red and showing cultivated land in light red and pink. The island's volcanoes show up as green-gray, and the recent lava flows are black.*

named Pu'u O'o. Lava from Kilauea flows through a natural volcanic tube for seven miles to emerge at Pu'u O'o and run down the face of the shoreline to fall into the ocean. On some days only steam issues from the vent, while on others red lava flows can be clearly seen, especially at night. The lava oozes like red-hot iron on the move. Kilauea has been erupting through Pu'u O'o since January 3, 1983, the longest eruption event ever recorded on Hawaii. It shows no signs of abating. By 1995 it had added five hundred acres of land to the island, and the flow continues at about 525,000 cubic yards per day.

Unlike Kilauea's constant eruptions, Mauna Loa, Hawaii's other active volcano, erupts only once about every four years. Mauna Loa stands 13,677 feet tall above sea level and is the largest active volcano in the world. Lava from this huge mountain covers about 50 percent of the island of Hawaii. Both Mauna Loa and Kilauea are broad shield volcanoes, the only type to exist in Hawaii; their volume consists of lava eruptions alone. The other type of volcano, the composite, is created from alternate eruptions of lava and ash. Mauna Loa is probably the tallest mountain on the earth, taller even than Mount Everest or K2. Most of the mountain is below the ocean waters, however. Altogether Mauna Loa rises 30,000 feet from the ocean floor and occupies a volume of 19,000 cubic miles and an area of 2,035 square miles.

Hawaii's volcanoes play a role in the divided climate of the island. On the east there is wet, moist weather that promotes the region's tropical vegetation, rivers, waterfalls and lush orchid farms. On the western side of the mountains the weather is dry and sunny, with beautiful sandy beaches, and coffee is grown on the slopes of the old volcanoes. Extinct volcanoes on the Island of Hawaii have become part of the landscape. The tallest mountain in Hawaii (measured above

314 Bottom
A fiery fountain at Pu'u O'o shoots lava into the air. When the lava quickly cools and falls back to earth, it can take in two forms. Long, glassy filaments that look like human hair or fiberglass are called Pele's Hair on Hawai'i, named for the

Hawaiian god of volcanoes. Little black droplets, which look like shiny pebbles, are called Pele's Tears.

314-315
The Mauna Loa eruption of 1984 was the last large eruption to date. Today, visitors can walk on trails

through the area of this lava flow to examine the now-cool volcano at close range. The region is considered to be a wilderness area by the National Park Service, and there are few amenities on the over forty-mile-long trail from the Kilauea area.

315 Bottom
The Pu'u O'o Vent of Kiluea is seen in this view, venting gases. Located on the flank of Kiluea, the vent is about seven miles from the volcano's caldera.

sea level only), 13,700 foot Mauna Kea, is a dormant volcano that dominates the scenery of the island. Rare plants on its upper reaches include the Hawaiian silversword, which looks like a delicate white yucca cactus. The flowering plant grows a tall greenish-yellow stalk with purple flowers as delicate as a feather in appearance. It is interesting to see such a delicate and subtly colored plant contrasted against the harsh black soil of the mountain.

In order to preserve the unique and incredible forces of volcanism and the rare species on Hawaii, the United States Congress established Hawaii Volcanoes National Park in 1916. About two-and-a-half million visitors a year come to Hawaii Volcanoes to see the awesome forces of nature which created the geography of the world as we know it today. Hawaii Volcanoes are a spectacular example of our world's timeless wonders.

Cover
Mount Fitzroy in the Patagonian Andes, Los Glaciares National Park, Argentina.
Photograph by Colin Monteath / Hedgehog House.

Front flap
Lake Powell, Utah, United States of America.
Photograph by Antonio Attini / Archivio White Star.

Back cover
Angel Falls, Venezuela.
Photograph by Silvestris.

Back flap
Landscape near Uchisar, Cappadocia, Turkey.
Photograph by Massimo Borchi / Archivio White Star.

Consultants: Natural history–Brian Cassie;
      Geology–Margaret Carruthers
Editor: Mary Beth Brewer
Designer: Clara Zanotti
Maps: Betty Vandone
Production Director: Louise Kurtz

First published in the United States of America in 2000 by Abbeville Press, 22 Cortlandt Street, New York, NY 10007

First published in Italy in 2000 by White Star S.r.l.
Copyright © 2000 White Star S.r.l. All rights reserved under

Library of Congress Cataloging-in-Publication Data

Moore, Robert J. (Robert John), 1956-
    Natural wonders of the world / [Robert J. Moore, Jr.]
      p.cm.
    ISBN 0-7892-0667-6
Title 1. Landforms. 2. Natural monuments. 3. Earth sciences. I

GB406 .M64 2000
551.41--dc21
                                                    00-032757